FUNDAMENTALS OF
Acid-Base Regulation

JAMES R. ROBINSON

M.D. Ph.D. Sc.D. (Cantab) F.R.S.N.Z. F.R.A.C.P.

Wolff Harris Professor and Chairman,
Department of Physiology, University of Otago
Medical School, Dunedin, New Zealand

FOURTH EDITION

BLACKWELL SCIENTIFIC PUBLICATIONS

LONDON OXFORD EDINBURGH MELBOURNE

Osney Mead, Oxford, England
3 Nottingham Street, London W1, England
9 Forrest Road, Edinburgh, Scotland
P.O. Box 9, North Balwyn, Victoria, Australia

ISBN 0 632 09200 9

First published 1961
Reprinted 1962
Second edition 1965
Third edition 1967
Reprinted 1969
Fourth edition 1972

Distributed in the U.S.A. by
F. A. Davis Company, 1915 Arch Street,
Philadelphia, Pennsylvania

Printed in Great Britain by
Holywell Press Ltd
Oxford
and bound at
Alden and Mowbray Ltd.
at the Alden Press, Oxford

CONTENTS

PREFACE TO THE FOURTH EDITION

Another new edition has been called for and again I have tried not to alter the character or purpose of the book. I have still discussed the regulation of reaction in terms of pH. After much thought and some consultation I decided not to adopt the International System (SI) of units completely as yet. Having renounced the pascal (the SI unit of pressure: 1 Pa = 1 Newton per square metre; 133.3 Pa = 1 mmHg) for partial pressures of gases I saw little point in abandoning the familiar mmHg in favour of the almost identical Torr (McGlashan, 1968; Aylward & Findlay, 1971). I have, somewhat reluctantly, abandoned equivalents (familiar to us but not to our children) in favour of mols, which our present students have heard of at school. I was strongly tempted to use mols per cubic metre instead of the old m-equiv/l, but have compromised with mmol/l, and will make the transition to $mol.m^{-3}$ in two steps if another edition is called for.

I want to thank the many favourable reviewers and others who helped to create the demand for a new edition; the smaller number of more critical reviewers and friends who will see that they helped to improve the book; and Mr Per Saugman without whose encouragement there would have been no book at all.

J. R. R.

DUNEDIN, NEW ZEALAND
April 1972

PREFACE TO THE FIRST EDITION

For some years I have been dissatisfied with my attempts to explain to medical students how the reaction of the body fluids is regulated. I suspect that such dissatisfaction is widespread, and there is a measure of agreement that part of the trouble arises from using the word 'base' in a sense which cannot be reconciled with the definitions of acids and bases in contemporary chemistry. Several authors have proposed more acceptable ways of approaching the problem, but their articles are in journals which most students cannot readily take away to brood over. (See Barker & Elkinton, 1958, for an appreciation of the situation with references.)

The effort to cast off the terminology with which we have become accustomed to confuse our understanding is all the harder because we are called upon to use familiar words with new meanings. But it turns out that when once we get used to the idea that the anions, not the cations, are the bases in the body's fluid media, a great deal of confusion and circumlocution can be avoided, and the things that happen in acid base metabolism can be described more naturally and simply. Moreover, there is no need to discard what has been written in the past, for 'translation' is easy.

Thus, the bicarbonate ion is a base in its own right; it is indeed a strong enough base to play a major part in keeping the blood slightly alkaline. Pitts, in deference to the clinical custom of regarding the metallic cations as bases, wrote of 'bicarbonate bound base'. But the modern theory of strong electrolytes treats salts in solution as completely dissociated into their ions, so that no particular cation can be thought of as 'bound' to any individual anion. Sodium bicarbonate when dissolved in water becomes 'sodium-and-bicarbonate', a mixture of positively charged sodium ions and negatively charged bicarbonate ions in equal numbers. The solution is alkaline because the bicarbonate ions take up hydrogen ions and lower their concentration. If we substitute 'bicarbonate' for 'bicarbonate bound base', Pitts' valuable papers (some of which are quoted as references) lose nothing of their clarity.

An important law of electrolyte solutions is the 'law of electroneutrality' which requires that the sum of the positive charges on the cations in any solution shall be equal to the sum of the negative charges on the anions. One consequence of this principle is that methods for determining the total concentration of the metallic cations (Na^+, K^+, Ca^{++} & Mg^{++}) in the plasma, which Peters & Van Slyke (1931) called the 'total base', do in fact measure the total concentration of base in the new sense; only they do so indirectly, by estimating the cations whose total concentration must be the same. Moreover, as Barker & Elkinton (1958) pointed out, the concept of the 'buffer base', which Singer & Hastings (1948) proposed to describe the acid base status of the blood, becomes even more valuable if the 'buffer base' is regarded as the total concentration of buffer anions, rather than as the concentration of the unspecified cations which balance their negative charges.

This small book is an attempt to describe how the kidneys, the lungs and the body's buffer systems co-operate to regulate acid base metabolism, using the words 'acid' and 'base' as chemists use them. The primary aim is to help the preclinical student and the interested clinician. Experts must be asked to accept this as some excuse for errors of judgment as to content or emphasis. The list of references is a guide to sources of more detailed information or discussion, and is not meant to be exhaustive. I have not ventured to deal with laboratory methods or with clinical and diagnostic details. The book sprang from the struggle to clarify my own ideas about the principles underlying the physiological regulation of reaction and its disturbances. Students and clinicians who first experienced this treatment of the subject in lectures led me to believe that the results of my struggle might be helpful to others. The real responsibility for their appearance in the present form belongs to Mr Per Saugman, of Blackwell Scientific Publications, to whom I am most grateful for his encouragement and for his material help in speeding an antipodean manuscript on its path to print.

J. R. R.

DUNEDIN, NEW ZEALAND
December 1960

PREFACE TO THE SECOND EDITION

In the preparation of this second edition little has been changed. Some points of detail have been amplified and a few references added. The account of respiratory control remains brief, but is a little less grossly oversimplified. The terminology has not been modified.

There is still argument about the best terms to use (see Campbell, 1962; Creese et al., 1962, and much correspondence in the *Lancet*), and the choice depends at least partly upon personal opinions about which terms are most convenient to handle or visualize. Campbell (1962) pointed out some disadvantages of the pH notation. But, apart from avoiding long strings of noughts, it has theoretical merit because, by the law of mass action, reacting species are influential in proportion to the products, not to the sums, of their concentrations. Equal steps on the (logarithmic) pH scale therefore represent equal increments (or decrements) of chemical potential; and for the present purpose pH has been retained.

I have still not attempted to give an account of diseases associated with disturbances of acid-base regulation, or of laboratory methods which can assist in the diagnosis and investigation of these diseases. This is partly because I am not competent to do so; and also because such additions would so lengthen the book as to destroy the brevity which must be one of its chief merits. It remains no more than an attempt to present the important underlying principles clearly.

Once again I want to express my thanks to Mr Per Saugman for the continuing interest and encouragement without which my contribution to this new edition would have been lacking. I also want to thank Professor R. A. McCance, C.B.E., F.R.S., for the hospitality of the Department of Experimental

Medicine where, during a period of Refresher Leave from the University of Otago in New Zealand, this revision was my first task.

J. R. R.

CAMBRIDGE, ENGLAND
March 1965

PREFACE TO THE THIRD EDITION

Although it is gratifying to be asked by the Publisher to prepare a new edition, the need for this one arose too soon and too urgently to permit as extensive a revision as might have been wished.

I have tried to deal more fairly with carbonic acid. A happier choice of some illustrative figures has simplified the tables. I have retained pH and have not altered my terminology despite some differences from the recommendations of an *ad hoc* Committee on Acid-Base Terminology (see Nahas, 1966). I still prefer to regard acidosis and alkalosis as abnormal states rather than as processes which could lead to abnormal states if they were not countered by secondary responses. The discussion of respiratory adjustments has been widened by reference to the cerebrospinal fluid and to the possible significance of intracranial chemoreceptors. I have included a little more about buffering in the tissues, and about renal acidosis, but have still deliberately omitted clinical and diagnostic matters for the reasons stated in the Preface to the Second Edition. The book has grown, but only by a few pages, and it retains its essential character and limitations.

Again I want to thank Mr Per Saugman, without whose encouragement neither this nor any earlier edition would have been written.

J. R. R.

DUNEDIN, NEW ZEALAND
February 1967

Chapter 1

INTRODUCTORY DEFINITIONS

Stability of reaction in the fluids of the body is part of Claude Bernard's 'fixité du milieu intérieur', and its importance is easy to see. Living systems consist largely of water; the biochemical reactions which supply their energy and their substance take place in solution, or at least at interfaces in contact with aqueous solutions; and the enzymes which catalyse those reactions require optimal acidity or alkalinity of the liquid phase in contact with them. As L. J. Henderson stated in 1913, 'Acidity and alkalinity surpass all other conditions, even temperature and the concentrations of reacting substances, in the influence which they exert upon many chemical processes.'

The reaction of a solution is acid when the concentration of hydrogen ions in it is greater than the concentration of hydroxyl ions; it is alkaline when the concentration of hydrogen ions is less than that of hydroxyl ions; and neutral when the concentrations of hydrogen ions and of hydroxyl ions are equal. These definitions should strictly be framed in terms of activities rather than of concentrations. Campbell (1968) nicely remarked that the activity measures what the concentration seems to be rather than what it actually is. Fortunately mammalian body fluids are dilute enough for concentrations to be used instead of activities without an important loss of accuracy.

The positively charged hydrogen ion is a proton, the atomic nucleus which remains after a hydrogen atom has lost its single electron. Protons cannot exist free in aqueous solutions, but combine with molecules of water. (Bell, 1952, stated that the proportion of protons likely to remain free is of the order of 1 in 10^{190}.) The hydrated proton in solution ought perhaps to be written as OH_3^+; it is sometimes referred to as the 'hydronium', 'oxonium' or 'hydroxonium' ion. For the present purpose, however, it will usually be sufficient to refer simply to 'hydrogen ions', and to denote these by the simpler and more familiar symbol 'H^+'. Square brackets will be used to denote concentrations; the concentration of hydrogen ions will be written as $[H^+]$, or else referred to as cH.

At any temperature, the product of the concentration of hydrogen ions and the concentration of hydroxyl ions in a solution in water is constant, so that the reaction of the solution can be defined precisely by specifying the concentration of hydrogen ions only. Except in extremely acid solutions the concentration of hydrogen ions is so small that it can only be expressed in moles per litre by inconveniently small numbers. Water at 25° C. contains only 0·0000001 mole of H^+ per litre. Campbell (1962) suggested that the writing of very small numbers could be avoided by expressing concentrations of H^+ ions in nanomoles per litre (1 nmol $= 10^{-9}$ mol, so that 1 mmol/l = 1,000,000 nmol/l). Concentrations can thus be expressed by convenient numbers which increase with acidity, and values of cH quoted in the following pages will be given in nmol/l unless it is stated otherwise. The pH notation with which most people have become familiar is almost as convenient for expressing very small concentrations, although it expresses them as *dilutions*, and on a logarithmic scale.

The pH of a solution may be thought of as the logarithm of the number of litres that would contain Avogadro's Number ($6{\cdot}023 \times 10^{23}$) or 1 gram of hydrogen ions. More strictly, pH represents the logarithm of the reciprocal of the activity of H^+ ions expressed in units of concentration. When electrometric methods are used to determine pH it is the activity of H^+ ions that is measured, and a knowledge of the activity coefficient for H^+ in a solution would be required in order to calculate the concentration of hydrogen ions from a measured value of pH. Activity and concentration are the same numerically only when the activity coefficient is unity, as it often is assumed to be in solutions as dilute as the fluids of the body. Illustrative values for cH in nmol/l quoted from time to time in the text are derived on the assumption that the activity coefficient is 1·00. In view of the difficulty of assigning values for the activity coefficients of single ions, concentrations of H^+ ions, even when they are derived from reliable measurements of pH, are all in some degree uncertain. In any event the concentrations in living systems are nearly always too small in comparison with those of the bulk ions to make any sensible contribution to ionic 'balance sheets' (e.g. 40 nmol/l in plasma compared with 4 mmol/l for K^+ and 145 mmol/l for Na^+). Moreover protons are physiologically active according to their chemical potential, which is a logarithmic function of their concentration, and is, of course, the variable to which a glass electrode responds (Waddell & Bates, 1969). pH has therefore been retained in preference to cH as the primary term in discussing the regulation of reaction. Further support for this choice will be found in an interesting essay by Davis (1967).

Chapter 2

THE NORMAL REACTION OF THE BODY FLUIDS

The reaction of the plasma of arterial blood in healthy persons corresponds to a concentration of H^+ ions of 0·00004 ± 0·000004 mmol or between 44 and 36 nmol per litre. In pH notation this is from pH 7·36 to pH 7·44. pH 7·40 is generally accepted as a convenient figure to remember for the average normal value. The range compatible with survival extends for about 0·4 pH unit on either side of the normal, say from 7·0 to 7·8 (Van Slyke, 1921), or maybe 6·85 to 7·65 (Gambino, 1966).

Since a difference of 1 unit of pH implies a tenfold change in concentration, the use of the logarithmic pH scale obscures the relative *insensitivity* of tissues to alterations in the concentration of hydrogen ions. We think of tissues as exquisitely sensitive because the reaction of the blood has to be controlled to a fraction of a unit of pH. But the range from pH 7·0 to pH 7·8 includes concentrations from 2½ times the normal (100 nmol per litre at pH 7·0) to two-fifths of the normal (16 nmol per litre at pH 7·8), i.e. a range from 40% to 250% of the value normally maintained. This is greater than the range of concentrations of potassium that can be tolerated, and far greater than the tolerable range of concentrations of sodium. (Sodium seems to be chiefly important because this cation and anions accompanying it account for most of the osmotic pressure of the extracellular fluids, and this osmotic pressure appears to be controlled with greater precision than any other physiological variable except temperature.) The greater tolerance of variations in the concentration of hydrogen than of other cations is not peculiar to man; most cells seem to be surprisingly resistant to changes in the external concentration of hydrogen ions (Bayliss, 1959–60; Vol. 2, p. 92).

At 25° C., the product of the concentrations in mol/l of hydrogen and hydroxyl ions in pure water is 10^{-14}, so that $[H^+] = [OH^-] = 10^{-7}$, and a pH of 7·00 corresponds to the neutral point at this temperature. The dissociation of pure water into equal numbers of hydrogen and hydroxyl ions increases as the temperature is raised, and at 38° C. the product $[H^+] . [OH^-]$ is no longer 10^{-14} but has increased to $10^{-13.6}$. Hence a neutral solution at the normal temperature of the body has a pH of 6·8 (cH 160), and the physiologically normal pH of 7·4 (cH 40) is somewhat more alkaline than has sometimes been supposed. The extracellular fluids were well described by Pitts as 'blandly alkaline' (Pitts et al., 1954). The clinical importance of maintaining this bland alkalinity is shown by the disturbances of function which accompany departures from it. Increasing alkalinity leads to headaches, mental confusion and lassitude accompanied by an increased neuro-muscular irritability which may present as tetany. Failure to maintain alkalinity leads to hyperventilation (with increase in depth more than in frequency of respiration) which may be so striking to the observer as to be called 'air hunger' though the patient may not be so aware of it. Impairment of central nervous function may be apparent as drowsiness passing through stupor to a terminal coma.

It is no small achievement for the body to maintain a reaction within the wider range of tolerance for survival, and especially within the narrower limits which permit normal function. Not only are large and inconstant amounts of acids produced in the course of metabolism, but, in addition, varying quantities of acids and also of alkalis are liable to be introduced into the body through the mouth.

The concentration of hydrogen ions inside the cells is likely to be more important for many biological activities than the reaction of the plasma and other extracellular fluids, but far less is known about it. The fluids inside the cells are probably less alkaline than the fluids outside, if not frankly acid. Measurements based upon the distribution of the weak acid 5,5-dimethyl-2,4-oxazolidine-dione (DMO), introduced by Waddell & Butler (1959) indicated that skeletal muscle in living rats had a pH of about 6·9 (Irvine et al., 1961), in agreement with earlier measurements of the distribution of bicarbonate (cf. Butler, 1966). Walker, Goodwin & Cohen (1969) extended the measurements to other organs and obtained rather similar results. (Carter, Rector & Seldin, 1966), however, using an intracellular glass microelectrode of novel design, recorded values around pH 6·0 in muscle cells of living rats. This is the value to be expected if H^+ ions were distributed in equilibrium with the electrical potential difference of about 90 mV across the muscle cell membrane. According to two valuable and complementary reviews (Robson, Bone & Lambie, 1968; Waddell & Bates, 1969) the majority support the higher value with its implication that H^+ ions do *not* reach an equilibrium distribution. This does not necessarily imply an active transport of protons because Conway (1957) pointed out that H^+ ions must be very slow to reach equilibrium because of their extremely low absolute concentration. There is moreover an unavoidable ambiguity in defining intracellular pH because of the great heterogeneity of cells, and discrepancies might arise not only from failure of equilibration but also because different methods measure different things. The DMO method may be expected to give an overall average for all phases of a vast number of cells, whereas a microelectrode may pick up the pH of free fluid in the tiny part of one cell that surrounds its tip. Whichever method is ultimately accepted as giving the correct result, there are problems enough for the future. Meanwhile it may be concluded that the fluids inside the cells are more acid than their surroundings; but it is not yet certain how great the difference is or how it is maintained. Under some conditions the reactions of intracellular and extracellular fluids may shift in opposite directions. This may occur, for example, when acid added to the blood stimulates respiration, which removes carbon dioxide from all the body fluids and so makes the cells more alkaline (Robin & Bromberg, 1959; see p. 89), or in potassium depletion, when the intracellular fluids tend to become more acid and the extracellular fluids more alkaline (Saunders et al., 1960; see p. 95).

Clinical assessment of disturbances in the mechanisms which regulate reaction depends mainly upon the analysis of plasma, which can be withdrawn to sample the extracellular fluid. There is moreover no doubt that the regulation of the reaction of the extracellular fluid is necessary for normal functioning of the cells, and may provide a background for the regulation of the reaction of the intracellular fluids by mechanisms yet largely to be discovered. The present discussion will be limited almost entirely to the reaction of extracellular fluids.

Chapter 3

'ACID-BASE BALANCE?'

The old term, 'Acid-base balance' had many disadvantages, for it rested upon a conception of the nature of acids and bases which is no longer acceptable to chemists. The cations Na^+, K^+, Ca^{++} and Mg^{++} were regarded as the 'bases' of the body's fluids. They were called 'fixed bases' because they were not volatile and could not be produced or destroyed by metabolic reactions, in contrast with ammonia (or ammonium) which was regarded as a volatile base capable of being produced or destroyed in metabolism.

'Anion-cation balance' is hardly a happier term than 'acid-base balance', for the condition of electroneutrality requires that the sum of the positive charges on the cations must precisely equal the sum of the negative charges on the anions in any solution. Guggenheim (1957, p. 372) calculated that an excess of univalent ions amounting to 10^{-10} mol in a region bounded by a spherical surface with a radius of 1 centimeter (a quantity far too small to be detected chemically) would be associated with a difference of potential of the order of 10 million volts. Hence in the most obvious sense, anion-cation balance cannot be disturbed but must always obtain in any solution. Except in gastric juice, however, the absolute concentration of H^+ ions is less than experimental errors in determination of the other ions, so that cH cannot be estimated by difference. Moreover the word 'balance' invites confusion with metabolic balance studies.

It seems preferable, therefore, to speak of the 'regulation of the reaction of the body fluids' rather than of 'anion-cation' or 'acid-base' balance. This avoids ambiguity, and the processes concerned can be described more straightforwardly if the terms 'acid' and 'base' are used as chemists now define them. The reaction of the body fluids is determined by the concentration (or activity) of hydrogen ions in them, and it will become clear that this in turn depends upon the proportions in which several acids and bases of different strengths are present.

Chapter 4

ACIDS AND BASES

In 1923 Brönsted in Denmark and Lowry in England suggested that an acid should be defined as a molecule or an ion with a tendency to give up a proton to the solvent; and that a base should be defined as a molecule or an ion with a tendency to take up a proton, and so to form an acid (Bell, 1952; Guggenheim, 1957). The more readily an acid gives up a proton, the stronger it is; and the more avidly a base takes up a proton, the stronger it is as a base. In water as solvent the reversible reaction

$$HB + H_2O \rightleftharpoons H_3O^+ + B^- \quad (1)$$

defines HB as an acid, and water as a base, accepting a proton to form H_3O^+. This reaction can be summarized as

$$HB \rightleftharpoons H^+ + B^- \quad (2)$$

which simultaneously defines HB as an acid and B^- as its conjugate base. The acid HB and its conjugate base B^- are known as a conjugate pair. It follows from the reversibility of the reaction (2) that the more readily HB gives up a proton ('the further the equilibrium of the reaction tends to be to the right'), the less readily does the conjugate base accept a proton. In other words, the stronger an acid, the weaker is its conjugate base; and the stronger a base, the weaker is its conjugate acid. This reciprocal relation between the strengths of the acid and base of a conjugate pair is illustrated by Table 1, which shows six acid-base pairs occurring in the body fluids.

Water appears twice in the Table; two molecules can interact thus:

$$2H_2O \rightleftharpoons OH_3^+ + OH^-. \quad (3)$$

Water is simultaneously a very weak base (it can take up a proton to form the very strong acid OH_3^+) and a very weak acid (it can give off a proton and form the very strong conjugate base OH^-). Both the very strong base and the very strong acid are formed in small amounts only (giving concentrations of the order of one mol in 10 million litres, as already stated), and pure water is neutral because these small amounts of acid and base are exactly equivalent.

TABLE 1. Examples of acid-base pairs (Strengths of acids indicated by approximate figures showing order of magnitude of pK′*).

	Order of pK′	*Acid*	$\rightleftharpoons$	*Proton + Conjugate Base*	
Very strong	−7	HCl	$\rightleftharpoons$	$H^+ + Cl^-$	Very weak
Very strong	−2	OH_3^+	$\rightleftharpoons$	$H^+ + H_2O$	Very weak
Fairly strong	4	H_2CO_3	$\rightleftharpoons$	$H^+ + HCO_3^-$	Rather weak
Rather weak	7	$H_2PO_4^-$	$\rightleftharpoons$	$H^+ + HPO_4^{2-}$	Fairly strong
Weaker	9·5	NH_4^+	$\rightleftharpoons$	$H^+ + NH_3$	Stronger
Very weak	16	H_2O	$\rightleftharpoons$	$H^+ + OH^-$	Very strong

Carbonic acid, here shown as a fairly strong acid with pK′ in the neighbourhood of 4 (3·6 according to Cotton & Wilkinson, 1962), has often been regarded as a weak acid with pK′ of the order of 6. But there is reason to believe that less than

*The pK′ is the logarithm of the reciprocal of an acid's dissociation constant (sometimes called the negative logarithm of the dissociation constant); the stronger an acid, the smaller is its pK′. The pK′ is equal to the pH of a dilute solution in which a weak acid is half neutralized, so that the concentrations of the acid and of its conjugate base are equal. The figures for strong acids cannot be interpreted literally in this manner; they still provide a scale of relative strengths determined in other ways. (See Bell, 1952.)

1% of carbon dioxide in aqueous solution is actually combined with the water as carbonic acid. Hence the concentration of carbonic acid, H_2CO_3, in a solution of carbon dioxide is very small, and the acid is proportionally stronger than might appear. The fact that sodium acetate in moist air smells of acetic acid would not be easy to explain if carbonic acid were as weak as has commonly been supposed, for acetic acid (pK′ about 4·7) would hardly be displaced by an acid of pK′ 6, although it should be displaced readily enough by an acid of pK′ 4.

The few examples in the Table illustrate that an acid may be either an uncharged molecule, a cation, or an anion. When a strong acid such as HCl is dissolved in water, the uncharged acid is so much stronger than OH_3^+ that it is almost completely converted into OH_3^+ and the weak conjugate base, Cl^-, thus:

$$HCl + H_2O \rightleftharpoons OH_3^+ + Cl^-, \qquad (4)$$

with the equilibrium very far indeed to the right.

The *bases* encountered in the body are either anions or uncharged molecules. It is clear that the cations cannot be called bases in the sense of the present definition. Indeed, on account of their positive charges, cations must always repel protons and could not be expected to unite with them. Neither can a cation like Na^+ give off a proton. The metallic cations are, in fact, neither acids nor bases. They neutralize the electrostatic charges upon the anions which accompany them, but make no direct contribution towards keeping the fluids of the body alkaline.

An aqueous solution of sodium hydroxide contains equal concentrations of the Na^+ ion, which is neither acidic nor basic, and of the strong base OH^-. Since the product $[H^+] . [OH^-]$ must remain constant, H^+ ions disappear from the solution by reacting with some of the added OH^- ions thus:

$$H^+ + OH^- \rightleftharpoons H_2O, \quad (5)$$

so that the concentration of hydrogen ions becomes less than that in pure water, and the solution is strongly alkaline because the concentration of hydrogen ions in it is far lower than the concentration of hydroxyl ions.

Chapter 5

BUFFERS IN THE FLUIDS OF THE BODY

We have seen that the alkalinity of the fluids of the body cannot be attributed to the cations which they contain. It is also evident that most of the anions in the body fluids, in so far as they are the conjugate bases of extremely strong acids, are such weak bases that they too can make no contribution towards maintaining a 'blandly alkaline' reaction. There are, however, besides Cl^- which is the predominant extracellular base, smaller but important concentrations of three stronger bases, bicarbonate (HCO_3^-), secondary phosphate (HPO_4^{2-}) and proteins. The proteins include plasma proteins and also haemoglobin and the proteins of the cells; physiologically normal reactions are on the alkaline side of their isoelectric points, so that they exist as polyvalent proteinate anions and may be symbolized as $Prot^{n-}$. The conjugate acids of these stronger bases are sufficiently weak for changes in pH in the neighbourhood of the neutral point to shift the equilibrium of the reaction

$$HB \rightleftharpoons H^+ + B^- \qquad (2)$$

to the left or to the right. If pH falls, the increasing concentration of hydrogen ions displaces the equilibrium towards the left and removes hydrogen ions from the solution. Conversely, if pH rises, the reduction in the concentration of hydrogen ions encourages the reaction to proceed to the right and add more hydrogen ions to the solution. In so far as these conjugate pairs take up hydrogen ions when the pH falls, and supply hydrogen ions when the pH rises, they stabilize the pH of any solution which contains them, and oppose alterations of reaction when strong acids or alkalis are added. For this reason they are known as buffers, and the anions may be conveniently called 'buffer bases'.

It will be useful to write out the reactions for the buffering of hydrogen ions by these buffer bases:

$$H^+ + HCO_3^- \rightarrow H_2CO_3(\rightarrow H_2O + CO_2); \quad (6)$$

$$H^+ + HPO_4^{2-} \rightarrow H_2PO_4^-; \quad (7)$$

$$H^+ + Prot^{n-} \rightarrow HProt^{(n-1)-}. \quad (8)$$

The strong acid OH_3^+ (written above as H^+) which is formed when an acid such as HCl is added to a solution containing one of these buffer bases, is replaced by the weak conjugate acid of the buffer base. The buffer bases behave as anions which either disappear from the solution altogether or suffer a reduction in their valency when a strong acid is added; they oppose the fall in pH by removing hydrogen ions at the same time. Hydrogen ions disappear from the solution by combining with the buffer bases, and the reaction may be written in the general form

$$H^+ + Buf^- \rightleftharpoons HBuf. \quad (9)$$

The negative charges on the buffer anions which disappear in reaction (9) are replaced by an equal number of negative charges on the extremely weak conjugate bases of strong acids which have been 'buffered' by the union of their hydrogen ions with the buffer bases. The sum of the negative charges on all the anions in the solution thus remains equal to the sum of the positive charges on the cations.

Chapter 6

WHY THE NORMAL pH IS 7·4

The blood is alkaline because it contains the quite strong bases HCO_3^-, HPO_4^{2-} and $Prot^{n-}$; but why should it be just so alkaline that its pH is 7·40? This may perhaps be seen most clearly by following up in detail an approach suggested by White, Handler & Smith (1968). A model of the plasma could be obtained by dissolving in a litre of water:

105 mmol of hydrochloric acid,
0·5 mmol of sulphuric acid,
6 mmol of the mixed organic acids of plasma,
2 mmol of phosphoric acid,
70 g approximately of plasma protein to supply 1 mmol of $Prot^{17-}$,
150 mmol of sodium hydroxide,
5 mmol of potassium hydroxide,
2·5 mmol of calcium chloride and
1 mmol of magnesium chloride.

This solution would resemble plasma in its concentrations of Na^+, K^+, Ca^{2+}, Mg^{2+}, Cl^-, SO_4^{2-}, phosphate and organic acid anions, all expressed per kg of water; but it would contain no bicarbonate and would be strongly alkaline because 24 mmol of hydroxyl ions would remain un-neutralized.

If this alkaline solution were exposed to an atmosphere in which the partial pressure of carbon dioxide was kept at 40 mmHg, it would absorb carbon dioxide; bicarbonate and hydrogen ions would be formed in equivalent quantities, thus:

$$H_2O + CO_2 \rightarrow H_2CO_3 \rightarrow H^+ + HCO_3^-. \qquad (10)$$

The hydrogen ions would unite with hydroxyl ions:

$$H^+ + OH^- \rightarrow H_2O, \tag{11}$$

and this would go on until 24 mmol of bicarbonate per kg of water had been formed to replace the initial excess of hydroxyl ions.

The solution would also contain dissolved carbon dioxide. The solubility of carbon dioxide in plasma at 37 °C. is 0·03p mmol per kg of water, where p is the partial pressure in mmHg (Woodbury, 1965). Hence if the partial pressure remained at 40 mmHg, the concentration of carbon dioxide in the solution after equilibrium had been attained would be approximately 40 × 0·03, which is 1·2 mmol/l. Most of this would be present as dissolved carbon dioxide in equilibrium with a far smaller concentration of carbonic acid. Berliner & Orloff (1956) gave the proportion of dissolved CO_2 combined to form H_2CO_3 as about 1 part in 900. More recent estimates (Bell, 1959; Edsall & Wyman, 1958) suggested that about 1 part in 300 is present as H_2CO_3 at 38 °C., and Chinard (1966) proposed an intermediate value of 1 part in 700. Thus although either would be described more accurately as containing 1·2 mmol/l of carbon dioxide, of which a small fraction is present as H_2CO_3, it is convenient to regard plasma or our model solution as containing 1·2 mmol/l of carbonic acid. This is indeed the amount of carbonic acid that would be determined by a titration carried out slowly enough to permit the hydration of the remaining carbon dioxide. On addition of alkali, the hydrogen ions from the small amount of H_2CO_3 initially present would be titrated rapidly, but carbon dioxide remaining in the solution would then react with water and form more H_2CO_3 to ionize and be titrated until all the carbon dioxide had been used up.

The concentrations of bicarbonate and of carbonic acid in a solution determine its reaction, as predicted by the Henderson-Hasselbalch equation, which may be written:

$$\mathrm{pH} = \mathrm{pK}' + \log \frac{[\text{Conjugate base}]}{[\text{Conjugate acid}]}. \qquad (12)$$

For a most interesting discussion of this famous equation see Homer Smith (1956), Appendix 2. The equation for the bicarbonate system is

$$\mathrm{pH} = 6{\cdot}1 + \log \frac{[HCO_3^-]}{[H_2CO_3]}. \qquad (13)$$

The value 6·1 is used for pK′ on the assumption that all carbon dioxide in the solution is present as a weak 'carbonic acid' with a pK′ of 6·1. This is a composite figure including the true pK′ of the molecular species H_2CO_3 and pK_h, which is equal to $-\log K_h$, where K_h is the equilibrium constant for the hydration of carbon dioxide and indicates the fraction of dissolved carbon dioxide which is combined with water as H_2CO_3. Thus 0·32% of H_2CO_3 with pK′ = 3·6 would give an apparent pK′ of 3·6 + log 100/0·32 = 3·6 + log 313 = 3·6 + 2·5 = 6·1.

A more useful form of the Henderson-Hasselbalch equation for plasma at the temperature of the body is

$$\mathrm{pH} = 6{\cdot}1 + \log \frac{[HCO_3^-]}{0{\cdot}03\, P\mathrm{CO}_2}. \qquad (14)$$

Equation (14) may be used to predict the pH of our model solution containing 24 mmol/l of bicarbonate and equilibrated with carbon dioxide at a partial pressure of 40 mmHg.

$$\mathrm{pH} = 6{\cdot}1 + \log \frac{[HCO_3^-]}{0{\cdot}03 \times 40}$$

$$= 6{\cdot}1 + \log \frac{24}{1{\cdot}2}$$

$$= 6{\cdot}1 + \log 20$$

$$= 6{\cdot}1 + 1{\cdot}3 = 7{\cdot}4$$

Equation (14) also shows clearly the two factors upon which the pH of the plasma principally depends. These are: the partial pressure of carbon dioxide, which determines the concentration of carbonic acid; and the concentration of bicarbonate, which is the difference between the total concentration of cations in the solution and the sum of the concentrations of all the anions other than bicarbonate. It will be convenient to illustrate the effects of these two factors separately, and then to consider their interaction.

Chapter 7

EFFECT OF VARYING PCO_2

It is easy to calculate the effect of the partial pressure of carbon dioxide upon the pH of a solution in which bicarbonate is the only buffer base present. Table 2 illustrates the application of equation 13 to a solution containing 24 mmol/l of bicarbonate. The solubility factor for CO_2 is taken to be 0·03 as in plasma. Every time the partial pressure of carbon dioxide is doubled, the pH should fall 0·3 unit; halving the partial pressure of carbon dioxide should raise the pH by 0·3 unit. The concentration of hydrogen ions in nmol/l turns out to be numerically the same as the partial pressure of carbon dioxide in mmHg. The consideration of a solution in which bicarbonate is the only buffer is not wholly artificial nor of merely academic interest, because bicarbonate is the only important buffer base in most extracellular fluids and is practically the only buffer base in the cerebrospinal fluid.

When other buffers are present besides bicarbonate, the situation is more complicated; the pH is lowered to a smaller

TABLE 2. Effect of PCO_2 upon pH of a solution containing 24 mmol of bicarbonate per litre at 37 °C

PCO_2	mmHg	10	20	40	80	160
0·03 PCO_2	mmol/l	0·3	0·6	1·2	2·4	4·8
$[HCO_3^-]$	mmol/l	24	24	24	24	24
$[HCO_3^-]$/0·03 PCO_2		80	40	20	10	5
log (buffer ratio)		1·9	1·6	1·3	1·0	0.7
pH		8·0	7.7	7.4	7·1	6·8
cH	nmol/l	10	20	40	80	160

extent by a given increase in the partial pressure of carbon dioxide, and raised less by a given decrease in $P\text{CO}_2$. Parsons (1920) compared the effect of carbon dioxide upon the pH of blood with the effect upon pH in solutions containing different concentrations of sodium bicarbonate. A given alteration in $P\text{CO}_2$ changed the pH of blood about two-thirds as much as it changed the pH of a solution containing a similar concentration of sodium bicarbonate. The reason for this smaller effect of carbon dioxide upon pH is that, when other buffers are present, the concentration of bicarbonate does not remain constant, but increases when the partial pressure of carbon dioxide is increased. As $P\text{CO}_2$ is increased, the plasma takes up carbon dioxide, more carbonic acid is formed, and the solution becomes more acid. The pH therefore falls towards the isoelectric points of the proteins in the plasma, and their valency as polyvalent anions is diminished. Similarly, as acidity increases, secondary phosphate becomes converted to primary phosphate, $HPO_4^{2-} \rightarrow H_2PO_4^-$, again with a reduction in the valency of the anion. Thus the number of equivalents of negative charge contributed by protein and phosphate (and by any further buffer anions other than bicarbonate which may be present) is diminished. In order that the sum of the negative charges on the anions shall remain equal to the sum of the positive charges on the cations in the system, additional anions are required. The balance of electrostatic charges is maintained by the formation of bicarbonate ions from the carbon dioxide taken up by the solution. As carbon dioxide is taken up, only a few of the hydrogen ions formed by the dissociation of newly-formed carbonic acid remain in the solution to lower its pH. Most of them disappear by combining with buffer bases which are present, for example,

$$H^+ + HPO_4^{2-} \rightarrow H_2PO_4^- \quad (7)$$

$$H^+ + \text{Prot}^{n-} \rightarrow \text{HProt}^{(n-1)-}. \quad (8)$$

Since all the new hydrogen ions which enter into these reactions with buffers are derived from carbon dioxide by the sequence of reactions:

$$H_2O + CO_2 \rightarrow H_2CO_3 \rightarrow H^+ + HCO_3^-, \qquad (10)$$

a new bicarbonate ion is formed for every hydrogen ion which combines with a buffer base other than bicarbonate. The hydrogen ions of course disappear when they combine with buffer anions, but the bicarbonate ions remain in the plasma to replace the $Prot^{n-}$ and the HPO_4^{2-} ions which have disappeared, thus keeping the sum of the negative charges on the anions equal to the sum of the positive charges on the cations in the system. The final result is that the total concentrations of cations and of anions remain the same as they were before the partial pressure of carbon dioxide was increased; but a greater fraction of the total concentration of anions is contributed by bicarbonate and a smaller fraction by other buffer bases such as protein and phosphate (Singer & Hastings, 1948; Elkinton & Danowski, 1955; Barker & Elkinton, 1958). The manner in which the concentration of bicarbonate is increased can be summed up and illustrated by combining equation (10),

$$H_2O + CO_2 \rightarrow H_2CO_3 \rightarrow H^+ + HCO_3^-,$$

with equation (9),

$$H^+ + Buf^- \rightarrow HBuf,$$

to give the overall result:

$$H_2O + CO_2 + Buf^- \rightarrow HBuf + HCO_3^-, \qquad (15)$$

where Buf^- now stands for buffer anions other than bicarbonate. Because the concentration of bicarbonate increases as well as the concentration of carbonic acid, the buffer ratio $[HCO_3^-] / [H_2CO_3]$ falls less than it would do if the partial pressure of carbon dioxide were similarly increased in a pure bicarbonate solution in which the concentration of bicarbonate remained almost constant. Since carbonic acid dissociates to give hydrogen and bicarbonate ions in equal numbers the concentration of bicarbonate could increase no more than the concentration of hydrogen ions on adding carbon dioxide to a solution buffered solely with bicarbonate.

Chapter 8

EFFECT OF VARYING THE CONCENTRATION OF BICARBONATE

Table 3 shows the effect of altering the concentration of bicarbonate in a solution containing bicarbonate as the only buffer in equilibrium with a fixed partial pressure of carbon dioxide. Doubling the concentration of bicarbonate raises the pH by 0·3 unit; halving the concentration of bicarbonate lowers the

TABLE 3. Effect of concentration of bicarbonate upon pH of a solution containing bicarbonate in equilibrium with 40 mmHg of carbon dioxide at 37 °C

PCO_2	mmHg	40	40	40
0·03 PCO_2	mmol/l	1·2	1·2	1·2
$[HCO_3^-]$	mmol/l	48	24	12
$[HCO_3^-]/0.03\ PCO_2$		40	20	10
log (buffer ratio)		1·6	1·3	1·0
pH		7·7	7·4	7·1
cH	nmol/l	20	40	80

pH by 0·3 unit. If we started with a litre of a solution containing 24 mmol of bicarbonate, kept in equilibrium all the time with a partial pressure of 40 mmHg of carbon dioxide, at 37 °C., and slowly added 24 mmol (24 ml of a normal solution) of sodium hydroxide, the pH should rise from 7·4 to 7·7. The addition of 12 mmol of HCl to a litre of the original solution should decompose one half of the bicarbonate which was present initially, and lower the pH from 7·4 to 7·1.

Table 3 does not accurately predict the changes in pH which would occur if similar quantities of acid and alkali were added to plasma or to blood. Other buffers present besides bicarbonate reduce the change in pH, for they share the buffering of added acid or alkali; and there is a smaller alteration in the concentration of bicarbonate. When acid is added to plasma, some hydrogen ions are taken up by other buffers, so that the amount of bicarbonate which disappears is less than the amount of acid added. When alkali is added, other buffers besides bicarbonate supply hydrogen ions to combine with the hydroxyl ions of the alkali; for example:

(*a*) $H_2PO_4^- \rightarrow HPO_4^{2-} + H^+; H^+ + OH^- \rightarrow H_2O;$
whence $H_2PO_4^- + OH^- \rightarrow HPO_4^{--} + H_2O.$ (16)

(*b*) $HProt^{(n-1)-} \rightarrow Prot^{n-} + H^+; H^+ + OH^- \rightarrow H_2O;$
whence $HProt^{(n-1)-} + OH^- \rightarrow Prot^{n-} + H_2O.$ (17)

Not all the hydrogen ions required to neutralize the added alkali have therefore to be supplied by carbonic acid; and fewer new bicarbonate ions are formed than if no other buffer was available. The increase in the concentration of bicarbonate is less than would be expected from the amount of alkali added to the extent that hydrogen ions are supplied by other buffer systems; and the increase in pH is correspondingly less.

Chapter 9

INTERACTION OF ALTERATIONS IN PCO_2 AND CONCENTRATION OF BICARBONATE

We have seen that when *either* the partial pressure of carbon dioxide *or* the concentration of bicarbonate is changed, while the other is kept fixed,

(i) altering $P\text{CO}_2$ changes pH in the opposite direction,

(ii) altering the concentration of bicarbonate changes pH in the same direction.

It should therefore be possible to compensate for alterations in pH resulting from changes in the concentration of bicarbonate by adjusting the partial pressure of carbon dioxide.

This is illustrated by Table 4, which again refers to a solution in which bicarbonate is the only buffer base. It will be seen that the increase of pH to 7·7 caused by adding 24 mmol of bicarbonate to a litre of solution containing 24 mmol of bicarbonate in equilibrium with carbon dioxide at a partial pressure of 40 mmHg at 37° C. (Table 3) could be compensated by raising $P\text{CO}_2$ to 80 mmHg. Similarly, the fall of pH to 7·1 caused by adding 12 mmol of strong monobasic acid (Table 3) could be compensated by lowering $P\text{CO}_2$ to 20 mmHg. So long as the partial pressure of carbon dioxide is made to vary in the same proportion as the concentration of bicarbonate, the buffer ratio remains at 20 and the pH remains at 7·4.

Compensation for the effect of adding acid by lowering the partial pressure of carbon dioxide has, however, an important limitation. Although the pH can be kept at 7·4 after half of

TABLE 4. Compensation for alteration in pH by changing PCO_2 to match concentration of bicarbonate in solutions of sodium bicarbonate at 37 °C

PCO_2	mmHg	80	20	10
$0{\cdot}03\ PCO_2$	mmol/l	2·4	0·6	0·3
$[HCO_3^-]$	mmol/l	48	12*	6*
$[HCO_3^-]/0.03\ PCO_2$		20	20	20
log (buffer ratio)		1·3	1·3	1·3
pH		7·4	7·4*	7·4*
cH	nmol/l	40	40	40

* pH is maintained; but notice depletion of buffer base.

the bicarbonate has been decomposed by the addition of acid, the capacity of the system to buffer further additions of acid has been halved with the concentration of bicarbonate. It would take only another 6 mmol of acid to bring the pH of a litre of the solution down to 7·1 if the partial pressure of carbon dioxide remained at 20 mmHg. The pH could again be restored to 7·4 by lowering the partial pressure of carbon dioxide still further, to 10 mmHg (Table 4), but the solution would then contain only 6 mmol of bicarbonate per litre, and 3 mmol of strong acid per litre would bring its pH down to 7·1 again if the partial pressure of carbon dioxide remained at 10 mmHg. The solution would still be very well buffered, in the sense that 3 mmol is 3,000,000 nmol of acid; and the fall in pH from 7·4 to 7·1 corresponds to an increase in cH from 40 to 80 nmol/l. Hence of 3,000,000 nmol of H^+ ions added to a litre of the solution, 40 nmol would remain free and the rest would be buffered, using up an equivalent amount of bicarbonate in the process.

Lowering the partial pressure of carbon dioxide can correct the change in pH which follows the addition of acid; but it does not prevent the reduction in buffering capacity as buffer base is used up. In order fully to restore the normal state, it is necessary, not only to return the pH to a normal value, but also to restore buffer capacity by replacing buffer base which has been used up. This distinction between compensation and repair was emphasized by Frazer and Stewart (1959).

Chapter 10

PHYSIOLOGICAL CONTROL OF REACTION

In so far as bicarbonate is the principal buffer base in mammalian extracellular fluids, the reaction of these fluids can be regulated by mechanisms which control the concentration of bicarbonate and the partial pressure of carbon dioxide. Both variables are subjected to physiological control, each by a separate system. The situation may be very briefly summarized thus:

(1) The partial pressure of carbon dioxide is maintained steady by the balance struck between the rate at which metabolic oxidations produce carbon dioxide and the rate at which the lungs remove it from the body. The rate of pulmonary ventilation is regulated by the respiratory centre so as normally to keep the partial pressure of carbon dioxide close to 40 mmHg.

(2) The concentration of bicarbonate is regulated by the kidneys. The tubular epithelium reabsorbs cations and anions from the glomerular filtrate in such a way that the sum of the concentrations of the cations in the plasma is normally kept greater than the sum of the concentrations of anions other than bicarbonate by about 24 to 27 mmol/l. The ready supply of carbon dioxide from metabolic sources ensures that the balance of anions required for electroneutrality is made up by bicarbonate (Pitts, 1950). The Henderson-Hasselbalch equation may therefore be crudely parodied for illustrative and mnemonic purposes:

$$\mathrm{pH} = \mathrm{pK} + \log \frac{\textit{Kidneys}}{\textit{Lungs and Respiratory Centre}}. \qquad (18)$$

The respiratory mechanism may now be discussed in a little more detail. The renal mechanism can be considered more conveniently after the buffering of acids and alkalis within the body has been dealt with.

Chapter 11

RESPIRATORY CONTROL OF REACTION

One of the most important functions of external respiration is the excretion of carbon dioxide, which is removed by the lungs at a rate proportional to the percentage of carbon dioxide in the alveolar gas and to the rate of alveolar ventilation. In a steady state of respiratory exchange, when carbon dioxide is removed as fast as it is produced, the percentage, and hence the partial pressure, of carbon dioxide in the alveolar gas must be inversely proportional to the rate of alveolar ventilation (see, for example, Carlson, 1960). Since, moreover, the partial pressure of carbon dioxide in the arterial blood corresponds to that in the alveolar gas, it follows that, for any rate of metabolic production, the arterial $P\text{CO}_2$ is determined by the rate of alveolar ventilation.

The ventilation of the lungs is in turn controlled by a number of factors, one of the most important of which is the partial pressure of carbon dioxide in the arterial blood perfusing the chemoreceptors and the respiratory centre. An increase in arterial $P\text{CO}_2$ acts through the respiratory centre to increase the rate of pulmonary ventilation, and a decrease in arterial $P\text{CO}_2$ reduces the rate of ventilation. Alterations in the partial pressure of carbon dioxide in the arterial blood are therefore quickly followed by changes in ventilation which tend to reverse them. This regulating mechanism is extremely sensitive: other things being equal, an increase in arterial $P\text{CO}_2$ to about 10 mmHg above the normal value of 40 mmHg increases the ventilation rate fourfold (Gray, 1949), and the rate of ventilation is correspondingly slowed if $P\text{CO}_2$ falls below the normal value. This great sensitivity of the respiratory mechanism ensures that the excretion of carbon dioxide keeps pace with its production, whilst at the same time the partial pressure of carbon dioxide in the arterial blood is kept close to 40 mmHg under resting conditions at sea level.

So long as the concentration of bicarbonate in the plasma is normal, a respiratory control which keeps the partial pressure of carbon dioxide close to 40 mmHg fixes the pH of the plasma at the same time. If, because acids or alkalis have been added to the blood, the concentration of bicarbonate is NOT normal, then the pH must depart from 7·4 unless the normal buffer ratio is restored by altering the partial pressure to re-adjust the concentration of carbonic acid to one-twentieth of the prevailing concentration of bicarbonate in the plasma. Such adjustments in fact occur, because the respiratory centre is influenced by the pH as well as by the $P\text{CO}_2$ of the arterial blood. Increasing acidity leads to an increased rate of pulmonary ventilation, whereas increased alkalinity reduces the rate of ventilation. The sensitivity of the respiratory centre is such that the ventilation rate is approximately doubled if pH falls 0·1 unit and halved if pH rises 0·1 unit, other things remaining unchanged (Gray, 1949). Acid added to the blood removes bicarbonate, but the acidity increases pulmonary ventilation and lowers the concentration of carbonic acid to match the low concentration of bicarbonate. Alkalis added to the blood increase the concentration of bicarbonate, but the rise in pH slows ventilation so that carbon dioxide is retained, and this brings about a parallel increase in the concentration of carbonic acid. In either case the respiratory mechanism stabilizes the buffer ratio by altering the rate of alveolar ventilation to match the concentration of carbonic acid to the concentration of bicarbonate in the plasma. When small quantities of acids or alkalis are ingested, variations in the rate of ventilation may be not obvious, but an increase in the percentage of carbon dioxide in the alveolar gas after the ingestion of sodium bicarbonate and a decrease following the ingestion of acids are more readily demonstrated. When large amounts of acid-forming salts are ingested, the increased ventilation may become obvious enough to resemble the 'air-hunger' which marks severe forms of acidosis encountered clinically in diabetic ketosis and in terminal uraemia. The descriptions of these phenomena in human subjects by Davies, Haldane and Kennaway (1920) and by Haldane (1921) are well worth reading (see also Haldane & Priestley, 1935).

The neurones which constitute the respiratory centre are subject to a wide variety of influences which modify their activity (see, for example, Asmussen, 1963). Besides $P\text{CO}_2$ they are influenced by the reaction and by the temperature of the blood which perfuses them. They can be driven voluntarily, and respiration may be affected by emotion. During physical exertion many factors, both neural and chemical, some still not completely understood, conspire to increase the responsiveness of the centre so that it shows a heightened sensitivity to CO_2, and ordinary values of arterial $P\text{CO}_2$ are associated with far greater rates of ventilation than would be expected under resting conditions. Indeed, the centre may be driven to regulate $P\text{CO}_2$ below resting levels. Anoxia directly depresses the respiratory neurones; but if the anoxia is not extreme, this depression is more than offset by impulses discharged from the aortic and carotid chemoreceptors, so that again the centre exhibits a lower threshold and an increased sensitivity to rising concentrations of CO_2. The arterial $P\text{CO}_2$ is accordingly regulated at levels considerably below 40 mmHg, and the blood tends to become more alkaline.

This description of the control of respiration by a respiratory centre and peripheral chemoreceptors in the carotid and aortic bodies is oversimplified. The respiratory response to carbon dioxide is mediated in part by intracranial chemosensitive elements which appear to be on or near the ventrolateral surface of the medulla medial to the roots of the ninth and tenth cranial nerves. This receptor area is bathed in cerebrospinal fluid, and it appears to stimulate respiration in response to increased activity of hydrogen ions in the cerebrospinal fluid, which is normally kept a little more acid than arterial plasma, with pH 7·32, cH 48 nmol/l (Mitchell, 1966). Pappenheimer (1967) and his group proposed that the receptive neurones are not on the surface and bathed in cerebrospinal fluid but in the interstitial fluid some little distance below the surface. Many outstanding problems and difficulties have been reviewed by Leusen (1972).

The intracranial chemoreceptors seem to modify respiration to regulate the reaction of the cerebrospinal fluid rather than that of the blood. When arterial $P\text{CO}_2$ increases carbon dioxide diffuses freely and the $P\text{CO}_2$ of the cerebrospinal fluid also increases, so that the ratio of H_2CO_3 to bicarbonate is increased and the cerebrospinal fluid becomes more acid. The intracranial receptors then reinforce the effects of stimulation of the peripheral chemoreceptors. When, however, the reaction of the blood is altered by addition of acid or alkali the concentration of bicarbonate in the cerebrospinal fluid does not change for some hours. During this time alterations in $P\text{CO}_2$ brought about by the respiratory response to stimulation of the peripheral chemoreceptors are transmitted to the cerebrospinal fluid, which therefore tends to become alkaline as the blood becomes acid, and vice versa. The intracranial chemoreceptors therefore act in opposition and moderate the response to the peripheral receptors. Later, changes in the concentration of bicarbonate, which may depend upon active secretion, correct the paradoxical alteration in reaction of the cerebrospinal fluid, and the intracranial receptors cease to oppose the peripheral ones. It thus transpires that the respiratory response to alterations in $P\text{CO}_2$, which reach the cerebrospinal fluid and threaten the reaction of the brain with a change in the same direction as that in the blood, is brisk, whereas the responses to the formation or absorption of acid or alkali, which would tend to shift the reactions of blood and brain in opposite directions, are more sluggish in onset; and they also tend to persist after the exciting causes have ceased to act.

Acclimatisation to different altitudes provides particularly interesting examples of interaction between the central and peripheral chemoreceptors. During the first few days at a high altitude the mechanisms controlling $P\text{O}_2$ and $P\text{CO}_2$ are in conflict. Pulmonary ventilation tends to be increased by hypoxic stimulation of the peripheral receptors, but the resulting reduction in $P\text{CO}_2$ makes the cerebrospinal fluid and the plasma more alkaline. The change is likely to be greater in the cerebrospinal fluid where bicarbonate is the only buffer base, and the intracranial receptors are mainly responsible for opposing the effects of hypoxia mediated by the peripheral receptors. The result is an inadequate supply of oxygen to the brain, and the syndrome of mountain sickness may appear. Over a few days the concentration of bicarbonate in the cerebrospinal fluid is reduced by unknown mechanisms which may include active transport (Severinghaus et al., 1963). This lowers the $P\text{CO}_2$ needed to stimulate the intracranial receptors and they cease to oppose the hypoxic drive from the peripheral receptors. Increased ventilation achieves a better supply of oxygen, but the low $P\text{CO}_2$ still leaves the plasma alkaline until, over a few more days, its concentration of bicarbonate is also reduced as the kidneys excrete bicarbonate in an alkaline urine in response to the lowered $P\text{CO}_2$ (Page 91). When the acclimatized person returns to sea level the hypoxic drive is suddenly removed; but any reduction in pulmonary ventilation would permit accumulation of CO_2 which would stimulate the peripheral receptors mildly and the central receptors more strongly by making the plasma and the cerebrospinal fluid more acid. This explains the curious observation that hyperventilation often continues for a day or so after returning to the lower altitude. It may be expected to continue until normal concentrations of bicarbonate have been restored, so that a $P\text{CO}_2$ of about 40 mmHg is once more required to maintain the normal reactions of cerebrospinal fluid and plasma.

A somewhat similar effect may be seen when a chronic metabolic acidosis is suddenly relieved, and hyperventilation continues until the low concentrations of bicarbonate which have developed in plasma and cerebrospinal fluid have been corrected. The possibility of conflict between the central and peripheral receptors is worth bearing in mind when laboratory reports on acutely ill patients reveal a discrepancy between ventilation, alveolar gas tensions and blood chemistry. The rate of ventilation, and the alveolar gas tensions to which it gives rise, may be determined by the composition of the cerebrospinal fluid rather than by that of the plasma.

Chapter 12

DISTURBANCES OF RESPIRATORY CONTROL

The normal respiratory compensation for alterations in the reaction of the blood can be disturbed by disease or overridden by experimental procedures. If the partial pressure of carbon dioxide ceases to be appropriate to the concentration of bicarbonate in the plasma, the buffer ratio deviates from 20, and the pH departs from 7·4, as illustrated in Table 2.

A. Hypoventilation

There is something to be gained by defining hypoventilation not merely as a reduced rate of pulmonary alveolar ventilation, but as a rate of pulmonary ventilation that is not adequate to remove carbon dioxide from the body as fast as it is produced without the assistance of an abnormally high percentage of carbon dioxide in the alveolar gas and a correspondingly increased partial pressure of carbon dioxide in the plasma. Hypoventilation may be caused by respiratory obstruction or paralysis, or by depression of the respiratory centre by drugs or toxins. Hypoventilation in this sense also occurs when the inspired air contains so much carbon dioxide that the ventilation rate cannot be increased enough to prevent the accumulation of carbon dioxide in the body. Although the rate of ventilation may then be far greater than normal, if it fails to prevent retention of carbon dioxide, it comes within the definition of hypoventilation. Hypoventilation is always associated with retention of carbon dioxide in the body, so that the concentration of carbonic acid is increased and the pH of the plasma falls, approximately in the manner illustrated in Table 2.

Some of the most extreme examples of the effects of hypoventilation have been found associated with respiratory failure when the concentration of carbon dioxide in the blood had become so high that it was acting as a narcotic and depressing instead of stimulating the respiratory centre. Wynn (1949) reported a $P\text{CO}_2$ of 197 mmHg in arterial blood, with pH lowered to 6·88, in a patient who ultimately recovered.

B. Hyperventilation

The term 'hyperventilation' is best used to describe a rate of pulmonary ventilation in excess of that required to remove carbon dioxide as fast as it is produced while the concentration in the alveolar gas is normal. Confusion may arise if the term is applied merely to an increased rate of pulmonary ventilation. The sense of the prefix 'hyper-' is not simply that the rate is increased, but that it is too great. If moderate exercise doubles the rate of production of carbon dioxide in the body, and the effective rate of alveolar ventilation is also doubled, carbon dioxide will be removed twice as fast as before, and the composition of the alveolar gas will not alter (Carlson, 1960). The new rate of ventilation is greater than the resting rate, but it cannot be called too great; an increase in ventilation which is appropriate to the greater need to remove carbon dioxide does not constitute hyperventilation. If the rate of pulmonary ventilation were quadrupled when the rate of production of carbon dioxide had only doubled, the increase in ventilation would be out of proportion to the need to remove carbon dioxide, and it would then be proper to speak of hyperventilation. Some degree of hyperventilation may indeed occur during violent exercise, for despite a greatly increased rate of production of carbon dioxide, the alveolar $P\text{CO}_2$ may be less than the value at rest, and this is good evidence that other factors besides the partial pressure of carbon dioxide contribute to the increased rate of ventilation which occurs during exercise.

Hyperventilation, in the restricted sense of a rate which lowers the partial pressure of carbon dioxide in the plasma, may occur through voluntary over-breathing for experimental purposes or to enable the breath to be held longer when swimming under water. Over-breathing also occurs in hysteria and as a response to emotional stress. Hyperventilation occurs at high altitudes when impulses set up in the chemoreceptors by blood poorly saturated with oxygen increase the rate of pulmonary ventilation. The rapid fall in $P\text{CO}_2$ in the cerebrospinal fluid at first reduces the contribution of the intracranial chemoreceptors to the respiratory drive and checks the hyperventilation until the concentration of bicarbonate in the cerebrospinal fluid is lowered; and the hyperventilation may persist after return to sea level until the concentration of bicarbonate has risen to its normal value.

Hyperventilation always removes carbon dioxide from the body and reduces the concentration of carbonic acid in the plasma, which becomes more alkaline. The partial pressure of carbon dioxide in the blood can be reduced to 10 to 15 mmHg by over-breathing. The resulting change in reaction is somewhat less than Table 2 indicates because of the presence of other buffers besides bicarbonate. Parsons' (1920) curves suggest that the pH of the blood might be increased by 0·2 to 0·3 pH unit, and such values have been observed (Brown, 1953). The greatly increased rate of ventilation which occurs during severe acidosis because of the effect of acidity upon the respiratory centre may be regarded as a compensatory hyperventilation which lowers the partial pressure of carbon dioxide and reduces the acidity of the plasma.

Chapter 13

DESCRIPTION OF DISTURBANCES

The disturbances of reaction caused by hypoventilation and by hyperventilation have usually been known as 'gaseous' or 'respiratory' acidosis and alkalosis. 'Gaseous' and 'respiratory' are appropriate epithets, for they indicate that the disturbances arise from mishandling of the blood gases or from an inappropriate rate of pulmonary ventilation. They thus emphasize a distinction from 'metabolic' disturbances of reaction, which arise from the excessive production or absorption of acids or alkalis, or from disorders of the secretion of acid by the renal tubular epithelium. A metabolic acidosis which occurs because the kidneys cannot excrete acids adequately may be called a 'renal acidosis'.

It has already been pointed out that when the reaction of a solution like plasma, which contains other buffers besides bicarbonate, is altered by a change in the partial pressure of carbon dioxide, there is no change in the total concentration of buffer base, but only a redistribution between bicarbonate and other buffer anions. The concentration of bicarbonate consequently increases when the partial pressure of carbon dioxide is increased, and decreases when the partial pressure of carbon dioxide diminishes. This leads to the somewhat paradoxical situation that the concentration of bicarbonate is increased in what has been known as 'respiratory acidosis', and is decreased in what has been known as 'respiratory alkalosis', Moreover, as will be explained in more detail in a later section, these primary changes, which are the immediate consequence of alterations in the partial pressure of carbon dioxide, are augmented by the action of the kidneys, which add more bicarbonate to the plasma when the partial pressure of carbon dioxide rises, but may excrete bicarbonate in the urine when the partial pressure of carbon dioxide falls.

It therefore seems preferable to restrict the terms 'acidosis' and 'alkalosis' to conditions in which the total concentration of buffer base is less than or greater than the normal. A primary respiratory disturbance, unaccompanied by any alteration in the total concentration of buffer base, is then neither acidosis or alkalosis. The cardinal manifestation of a primary respiratory disturbance is an alteration in the reaction of the blood, which may appropriately be called 'acidaemia' or 'alkalaemia'. The four terms required to describe disturbances of reaction and of the mechanisms which maintain reaction may then be defined as follows:

ACIDOSIS. A condition in which the total concentration of buffer base is less than normal.

ALKALOSIS. A condition in which the total concentration of buffer base is greater than normal.

ACIDAEMIA. An abnormally acid reaction of the blood (plasma pH less than 7·36).

ALKALAEMIA. An abnormally alkaline reaction of the blood (plasma pH greater than 7·44).

Various combinations of these four primary conditions are possible. Acidosis may be accompanied by acidaemia, but need not be so, for the change in pH may be prevented by respiratory removal of carbon dioxide. Similarly, alkalosis will not be accompanied by alkalaemia if enough carbon dioxide has been retained to prevent the change in reaction. When the kidneys respond to a rising partial pressure of carbon dioxide by increasing the reabsorption of bicarbonate, they create a renal alkalosis of metabolic type to counteract a gaseous acidaemia. The fact that the total concentration of buffer base is thereby ultimately *increased* makes 'respiratory acidosis' a confusing name for the complex disturbance; 'respiratory acidaemia' is less confusing. In the initial stages of an excessively rapid excretion of carbon dioxide by the lungs, there is neither acidosis nor alkalosis, but only alkalaemia. The action of the kidneys when they respond to the falling partial pressure of CO_2 in their arterial blood by excreting bicarbonate in the urine (p. 91) is to reduce the concentration of bicarbonate in the plasma and so to create a renal *acidosis*. The fact that the concentration of buffer base in the blood is thereby reduced makes 'respiratory alkalosis' an inappropriate name for this condition. The disturbance which is encountered at high altitudes when pulmonary ventilation is increased in response to hypoxia is better called 'respiratory alkalaemia'. These points are illustrated in Fig. 1, a block diagram which represents a modification and extension of Fig. 10·4 from the book by Elkinton & Danowski (1955).

The first column of Fig. 1 depicts the normal state of whole blood. The total buffer base (BB) is shown as made up of two components, bicarbonate and 'protein-and-phosphate'. pH depends upon the ratio of the concentration of bicarbonate to the concentration of 'carbonic acid', which is shown above the thick horizontal line on a different scale.

Column 2 shows the immediate effect of hypoventilation. Total concentration of buffer base is not altered, but the fall in pH is accompanied by a decrease in protein-and-phosphate and an increase in bicarbonate. The increase in bicarbonate is quite insufficient to prevent the fall of pH, for the increase in P_{CO_2}, and the corresponding increase in concentration of carbonic acid, are proportionally far greater. The kidneys later act to increase the concentration of bicarbonate (Column 3) so that the total buffer base is actually increased. The appropriate name for the respiratory disturbance is clearly 'acidaemia' rather than 'acidosis'.

Column 4 shows the immediate effect of hyperventilation.

The total concentration of buffer base is unchanged, but pH has risen. Protein and phosphate have given up hydrogen ions, so that their valency is increased and they make a greater contribution to the sum of anionic charges. The contribution of bicarbonate is correspondingly less, but its concentration is reduced far less in proportion than that of carbonic acid. The kidneys later respond by excreting bicarbonate in the urine, so that the concentration of bicarbonate falls further (Column 5) and the total concentration of buffer base is actually decreased. The appropriate name is therefore 'alkalaemia' rather than 'alkalosis'.

Column 6 depicts the result which might be expected to follow the ingestion of ammonium chloride, if respiratory compensation did not occur; $P\text{CO}_2$ is still shown as 40 mmHg. Bicarbonate has been displaced from the plasma by chloride, and the fall in pH has also been accompanied by a reduction in the concentration of other buffer anions by uptake of hydrogen ions. Total buffer base is thus diminished, and the appropriate name for the disturbance is 'acidosis'. There is also acidaemia; but the response of the respiratory centre to the acid reaction of the plasma lowers $P\text{CO}_2$ to match the low concentration of bicarbonate, so that the buffer ratio and the pH are returned towards normal values. The acidaemia is partly corrected, but the acidosis remains (Column 7). The acidosis itself will be more slowly corrected as the kidneys excrete hydrogen ions (both as titratable acid, and combined with ammonia as ammonium—see later, p. 77), together with the excess chloride, and restore bicarbonate to the plasma.

Column 8 shows the effects of ingesting sodium bicarbonate, again as they might be expected to appear if no respiratory compensation occurred. The concentrations of bicarbonate and of total buffer anions are increased, and the condition is clearly an alkalosis. The accompanying alkalaemia slows pulmonary ventilation so that $P\text{CO}_2$ increases (Column 9) to match the high concentration of bicarbonate. The alkalaemia is thus partly corrected, but the alkalosis remains until the kidneys have had time to excrete the excess of bicarbonate in an alkaline urine.

The relations illustrated in Fig. 1 may be summed up as follows. The essential feature of a metabolic disturbance is an increase or a decrease in the total concentration of buffer base, denoted by the term alkalosis or acidosis; the change of pH tends to be small because of respiratory compensation. Disorders of ventilation lead to rather pure disturbances of pH without primarily affecting the total concentration of buffer base, and are best described as acidaemia and alkalaemia.

The unambiguous assessment of disturbances of acid base metabolism requires a knowledge of the pH of the plasma. In principle this can be arrived at indirectly through the Henderson-Hasselbalch equation (p. 18) if $[HCO_3^-]$ and P_{CO_2} are both known; but it is usually more convenient to measure pH than to determine P_{CO_2}. The concentration of bicarbonate can be derived from the volume of carbon dioxide given off when plasma is treated with acid and exposed to a vacuum. Thus 20/21 of the total carbon dioxide comes from bicarbonate if the plasma was initially at pH 7·4; since at STP 22·3 ml of CO_2 is one mmol, and is equivalent to 1 mmol of bicarbonate, it follows that 1 ml of carbon dioxide per 100 ml of plasma corresponds to $\frac{20}{21} \times \frac{10}{22{\cdot}3}$ or 0·426 mmol of bicarbonate per litre. The factor depends upon the buffer ratio, and it is a salutary exercise to work out the extent of its dependence upon pH. The concentration of bicarbonate is not, however, a sufficient indication in itself, for the plasma contains other buffer bases as well as bicarbonate. Moreover it can be misleading, because the concentration of bicarbonate is increased in respiratory acidaemia; and it is decreased in respiratory alkalaemia as well as in metabolic acidosis (Fig. 1). Both

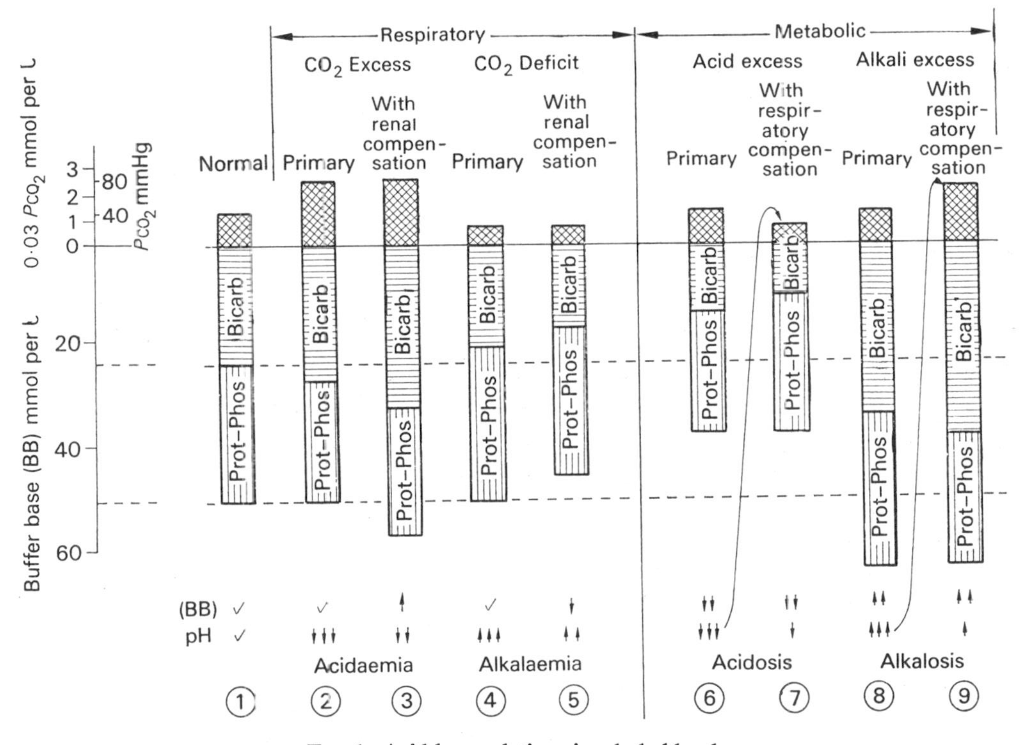

FIG. 1. Acid-base relations in whole blood.

respiratory alkalaemia and metabolic acidosis are associated with increased rates of pulmonary ventilation, but the pH of the plasma is low in metabolic acidosis. *If metabolic acidosis is known to exist*, the falling concentration of bicarbonate as the plasma takes up acid reflects the dwindling reserve of the most important buffer base which is available to deal with further quantities of acid. There is then some justification for referring to the concentration of bicarbonate as 'the alkali reserve'; but it may be preferable to stop using the term in order to avoid confusion from its paradoxical behaviour in respiratory disturbances of reaction. Astrup (1961) and his colleagues introduced a method whereby a 'standard bicarbonate' concentration (referred to plasma of whole blood with its haemoglobin saturated with oxygen at 38 °C.) can be determined from a nomogram after measuring the pH of a sample of blood initially and also after equilibration with 4% and with 8% CO_2 in oxygen (Siggaard Andersen et al., 1960). Methods for the clinical assessment of disordered acid-base metabolism are, however, beyond the scope of this account. For further discussion and references, see Singer & Hastings (1948), Elkinton & Danowski (1955), and Davenport (1958). Campbell (1963) provided a valuable and concise account of this topic. A comprehensive contemporary review of acid-base measurement has been issued by the New York Academy of Sciences (Nahas, 1966). The authors did not generally favour the definitions of acidosis and alkalosis on page 41 above; they preferred to define these as processes tending to alter pH rather than as alterations in the composition of body fluids. Useful comments on terminology and its historical development will also be found in a book by Filley (1971).

Chapter 14

METABOLIC PRODUCTION OF ACID AND ITS DISPOSAL

A. Carbonic Acid

Carbonic acid is produced by metabolic reactions in far greater quantity than any other acid. Organic acids which occur in sequences of metabolic reactions are ultimately oxidized to yield carbon dioxide and water, which form carbonic acid. A man produces about 13,000 mmol of carbon dioxide every day, nearly 10 mmol/min. Although H_2CO_3 could yield two hydrogen ions, only a tiny fraction is ionized as carbonate, CO_3^{2-}, in the body fluids, and 13 mol of sodium hydroxide would be required to convert the day's carbon dioxide to bicarbonate at pH 7·4. The daily output of carbon dioxide is thus equivalent in neutralizing power to 13 mol of hydrochloric acid, or over a litre of the fuming, concentrated hydrochloric acid which is about 10 M.

Fortunately, carbonic acid is readily converted to the volatile anhydride, carbon dioxide, which can be excreted by the lungs. Fortunately, too, the small extent to which carbon dioxide combines with water to form H_2CO_3 makes carbonic acid behave as a weaker acid than it truly is, so that its conjugate base is strong enough to serve as one of the principal buffer bases in the body fluids. Not only, then, does the large amount of waste acid arising from metabolism cause little embarrassment; it is even exploited as a means of controlling the reaction of the body's fluids.

The erythrocytes play an important part in transporting and buffering carbon dioxide. They contain the enzyme, carbonic anhydrase, which enables the bicarbonate of the plasma to be converted, through carbonic acid, to water and free carbon dioxide quickly enough for the carbon dioxide to be unloaded during the short time which the blood spends in the pulmonary capillaries.

The erythrocytes also contain a large concentration of the remarkable protein, haemoglobin, which has other important properties in addition to its well-known ability to combine loosely and reversibly with oxygen. It carries some of the carbon dioxide in the whole blood reversibly bound as a carbamino compound; and although this only amounts to about 5% of the total carbon dioxide of the blood, it may be as much as a third of the carbon dioxide which is taken up rapidly when the blood is in the tissues and given up rapidly when the blood is in the lungs. Moreover, the uptake of carbon dioxide to form carbaminohaemoglobin in the tissues and the unloading of carbamino CO_2 in the lungs are largely effected by the alternate reduction and oxygenation of haemoglobin. At a P_{CO_2} of 40 mmHg oxyhaemoglobin holds 0·1 and reduced haemoglobin a little more than 0·3 mmol of carbamino CO_2 per mmol of haemoglobin. The effect of P_{CO_2} itself is relatively trivial.

The formation of carbaminohaemoglobin cannot be described as a process of buffering, because one hydrogen ion is released for each molecule of CO_2 taken up (Davenport, 1958). Haemoglobin is, however, after bicarbonate itself, the most important buffer in the whole blood. Its peculiar importance as a buffer for carbonic acid depends upon the effect of oxygenation upon its readiness to take up or give off hydrogen ions, together with the fact that the amounts of oxygen and carbon dioxide in the blood usually vary simultaneously in opposite directions. Reduced haemoglobin accepts hydrogen ions more readily than oxyhaemoglobin, and is therefore a stronger base; or, as it is more often stated, oxyhaemoglobin is a stronger acid than reduced haemoglobin. When the blood is passing through the tissues and taking up carbon dioxide, it is also losing oxygen so that oxyhaemoglobin is being replaced by reduced haemoglobin. The increased acidity which would be expected to follow the formation of carbonic acid is almost precisely neutralized by the reduction of oxyhaemoglobin to a stronger base which takes up hydrogen ions as they are released from carbonic acid. In fact haemoglobin is so much a weaker acid than oxyhaemoglobin that unless 0·7 mmol of H^+ ions are supplied for every mmol of oxyhaemoglobin which is reduced, the blood actually becomes more alkaline (Davenport, 1958). And when the blood is passing through the lungs, it tends to become acid unless 0·7 mmol of hydrogen ions are removed for every mmol of oxyhaemoglobin that is formed.

If the respiratory exchange ratio is 0·7, 0·7 mmol of carbon dioxide will be given up to the alveolar gas for each mmol of oxygen taken up by haemoglobin; and the 0·7 mmol of H^+ ions made available by oxygenation of haemoglobin will release 0·7 mmol of carbon dioxide from bicarbonate in the plasma:

$$\underset{0\cdot7\ \text{mmol}}{HCO_3^-} + H^+ \rightarrow \underset{0\cdot7\ \text{mmol}}{H_2CO_3} \rightarrow H_2O + \underset{0\cdot7\ \text{mmol}}{CO_2} \qquad (6)$$

Consequently the formation of one mmol of oxyhaemoglobin and the loss of 0·7 mmol of carbon dioxide can occur without any change in pH. Similarly, in the tissues, if 0·7 mmol of carbon dioxide is formed for each mmol of oxygen utilized, the hydrogen ions of the carbonic acid formed by hydration of the carbon dioxide will just suffice to prevent a rise of pH which would otherwise occur when 1 mmol of oxyhaemoglobin gave up its oxygen.

Thus, if the respiratory exchange ratio is 0·7, all hydrogen ions derived from carbon dioxide generated in metabolism can be buffered by haemoglobin alone without any alteration in pH. The respiratory exchange ratio is not usually equal to 0·7, but somewhat greater. In this case the other buffers of the blood are only required to deal with (R–0·7) mmol of hydrogen ions for each mmol (22·4 ml) of oxygen consumed in the tissues.

This buffering by haemoglobin of hydrogen ions formed from carbon dioxide in the blood may be summarized in the following equations:

$$CO_2 + H_2O \text{ —(carbonic anhydrase)→ } H_2CO_3,$$

$$H_2CO_3 \rightarrow HCO_3^- + H^+, \quad (19)$$

$$O_2Hb^- + H^+ \rightarrow HHb + O_2. \quad (20)$$

Adding these together, the final result is:

$$O_2Hb^- + CO_2 + H_2O \rightarrow HHb + O_2 + HCO_3^-. \quad (21)$$

The hydrogen ions are taken up by haemoglobin and remain in the erythrocytes, but the bicarbonate moves out into the plasma, partly in exchange for chloride (chloride shift). Most of the carbon dioxide added to the blood therefore appears, not as *acid*, at all, but as bicarbonate in the plasma. It is the presence of both haemoglobin and carbonic anhydrase inside them which enables the erythrocytes to perform this important little conjuring trick.

The carriage and exchange of the blood gases are dealt with in standard textbooks of physiology. The quantitative aspects are set out with great lucidity in detail by Davenport (1958).

B. Other Acids

A few strong acids are generated by metabolic reactions in quantities which are small compared with that of carbonic acid (less than 100 mmol daily, compared with a potential 13,000), but which are important because these acids are not ʌolatile and have to be excreted by the kidneys.

No significant amounts of organic acids require to be eliminated in this way under normal conditions, but during severe diabetic ketosis the so-called 'acetone bodies', principally β-hydroxybutyric and acetoacetic acids, are produced at rates which may be as great as 1000 mmol per day.

The most important acid product of normal metabolism is the 'mineral acid', sulphuric acid, which is formed by oxidation from sulphur in the amino acids methionine and cysteine. According to Homer Smith (1951), the metabolism of 100 grams of protein leaves 30 mmol of sulphuric acid to be disposed of. For many years phosphoric acid has been classed with sulphuric acid as a mineral acid formed during metabolism, and Smith (1951) added that the metabolism of 100 grams of protein also left an amount of phosphoric acid which would require 50 mmol of sodium hydroxide to bring it to pH 7·4; and, further, that the metabolism of 100 grams of fat containing 10% of lecithin would release another 50 mmol of phosphoric acid. Hunt (1956) pointed out that phosphoric acid is ingested preformed in the phosphate groups of phospholipids, phosphoproteins and phosphoric esters; it is merely released in the body, not formed by oxidation as sulphuric acid is formed from sulphur ingested in other, less highly oxidized forms. Richet et al. (1967), however, claimed that up to 15 mmol of H^+ ion per day on ordinary diets was generated by metabolism of phospholipids. Both Hunt (1956) and Lemann and Relman (1959) were able to show that the amount of acid formed in the body and excreted in the urine of human subjects was principally determined by the amount of sulphur metabolized. As will be seen in a later section (p. 77), most of the titratable aid of the urine is present in the form of primary phosphate ($H_2PO_4^-$) formed by coupling H^+ ions secreted by the tubular epithelium with secondary phosphate (HPO_4^{2-}) from the glomerular filtrate. The relation between the sulphuric acid which is generated and the phosphoric acid which is released in metabolism is thus probably that the phosphate is used as a supply of buffer with which to excrete some of the hydrogen ions from the sulphuric acid. Sulphuric acid itself is far too strong to be excreted as free acid (with hydrogen ions balancing the negative charges on the sulphate anion, SO_4^{2-}) in the urine, which cannot be made more acid than about pH 4·5. Other hydrogen ions derived from the sulphuric acid formed in the body are excreted in the urine combined with ammonia to form ammonium, NH_4^+, a weak acid which is also a cation. The important point is that the total amount of acid excreted in the two forms of titratable acid and ammonium corresponds more closely to the amount of sulphur metabolized than to the total amount of sulphate and phosphate. It is between 50 and 100 mmol daily on an ordinary mixed diet, and the amount may be increased by the ingestion of methionine.

The ingestion of ammonium chloride also increases the amount of acid formed in the body and excreted in the urine. After ammonium chloride has been absorbed, the ammonium ions become split into ammonia, which is removed by conversion to urea in the liver, and the hydrogen ions, which are left behind with the chloride ions. The final effect is equivalent to that of ingesting hydrochloric acid.

$$Cl^- + NH_4^+ \rightarrow NH_3 \text{ (removed)} + H^+ + Cl^-. \quad (22)$$

Salts of the poorly absorbed divalent alkaline earth cations also increase the acidity of the body fluids. It is supposed that when, for example, calcium chloride is ingested, the chloride is absorbed from the intestine more rapidly than the calcium. Some of the calcium remains in the faeces combined with phosphate and bicarbonate which have been withdrawn from the plasma, and the plasma is made acid by losing bicarbonate in exchange for chloride.

Alkalinizing salts, such as potassium citrate and sodium

lactate, as well as vegetable diets which contain large amounts of potassium balanced by organic anions, reduce the net amount of acid formed in the body and may cause the urine to become alkaline. The organic anions which are absorbed are apparently metabolized, leaving bicarbonate behind to accompany the cations, so that the ingestion and subsequent metabolism of these salts is tantamount to the ingestion of bicarbonate.

The renal excretion of acids (of which the mechanism will be considered in a later section) is slow compared with the excretion of carbonic acid by the lungs, so that strongly acid waste products of metabolism have to be retained in the body for considerable periods of time. This must be done without grossly altering the reaction of the body fluids, and it will now be convenient to outline the mechanisms which are available to buffer acid products of metabolism or ingested acids or alkalis while these remain in the body and are transported from their sites of production or absorption to the kidneys to be eliminated.

Chapter 15

HANDLING OF ACIDS AND ALKALIS WITHIN THE BODY

A. Dilution and Metabolic Transformation

Acid and alkaline substances which tend to disturb the reaction of the body fluids do not appear simultaneously throughout those fluids. They are produced at specific points inside cells, or absorbed from relatively restricted portions of the alimentary tract. Thus lactic acid is produced by enzymes at specific sites in the cells of active muscles. Much of this lactic acid is oxidized, ultimately to carbon dioxide and water, or else rebuilt into glycogen. The remainder, which escapes metabolic transformation, is first diluted in the whole fluid contents of the cells in which it is produced, and then diffuses out of these cells and is diluted throughout the extracellular fluids of the body. Some enters other cells, where it is diluted in their contents, and, in some instances, is destroyed by oxidation or other metabolic change. Sulphuric acid from the oxidation of sulphur-containing amino acids may be diluted in all the fluids of the body like lactic acid. It is not subject to metabolic degradation, but some sulphuric acid is conjugated with organic molecules and appears in the urine as neutral 'ethereal' sulphate.

The dilution of an acid or alkaline product of metabolism in the whole mass of the body fluids is the simplest of the several devices which enable such substances to be held temporarily in the body with a minimal alteration in the reaction of the blood and other fluids. Each tenfold dilution of a strong acid raises the pH by 1 unit. If, however, all of the 50 mmol or so of strong acid produced by a day's metabolic activity were diluted in the body's total content of about 50 litres of water, the final concentration would be about 1 mmol/l; and the pH of a solution containing 1 mmol of strong acid per litre should be 3·0. Important though it is as an immediate means of preventing intense local changes in reaction, dilution alone is quite inadequate to deal with the acid products of metabolism in a medium which cannot be allowed to become more acid than pH 7·00.

B. Buffering

A much more effective protection against the effects of non-neutral and relatively indestructible products of metabolism is provided by the buffers which are present in the several phases of the body's fluids. The buffering capacity of a living animal's tissues was beautifully demonstrated by Pitts (1953), who compared the change of pH produced by infusing hydrochloric acid into a dog with the change of pH when the same amount of hydrochloric acid was added to as much water as the dog's body contained. 156 mmol of HCl reduced the pH of the arterial blood of a dog weighing 19·7 kg from 7·44 to 7·14. When 156 mmol of HCl was added to 11 litres of water the pH fell to 1·84. The difference between the two final results may be attributed mainly to the buffers in the dog's tissues and body fluids. Although emphasis was formerly placed upon the buffers in the blood, because of their greater accessibility to investigation, it is now realized that these buffers are supplemented both by buffering within the cells of the tissues and by an exchange of ions between the extracellular fluids and the mineral constituents of the skeleton.

(i) *Buffers in the Blood*

The most important buffers in the blood are the haemoglobin in the erythrocytes, and the bicarbonate, protein and phosphate in the plasma. Table 5 illustrates their relative contributions to the buffering of an amount of acid which would lower the pH of 1 litre of whole blood from 7·4 to 7·0 (Peters & Van Slyke, 1931). Buffers of which a constant amount is present are most effective in a range of pH close to the pK′ of the buffer acid, when the buffer ratio is not far removed from 1·0.

Bicarbonate appears in Table 5 as an astonishingly valuable buffer despite an extremely unfavourable buffer ratio. This ratio is about 20 : 1, the same as the ratio of [HCO_3^-] to 0·03 P_{CO_2}, if all dissolved carbon dioxide is supposed to be present as a weak 'carbonic acid' of pK′ 6·1. But the buffer ratio [HCO_3^-] : [H_2CO_3] must be of the order of 10,000 : 1 for the quite strongly acid molecular species H_2CO_3 with its

TABLE 5. mmol of hydrogen ions taken up by individual buffers when 1 litre of blood is titrated from pH 7·4 to pH 7·0

Bicarbonate	18
Phosphate	0·3
Plasma protein	1·7
Haemoglobin	8
Total	28

pK′ less than 4. The carbonic acid-bicarbonate pair would be a poor buffer indeed in a closed system. Its great capacity as a physiological buffer depends upon the relatively large concentration that is present (24 mmol/l) and also upon the ease with which the body can alter not the valency, but the amount of buffer anion present. The body is, of course, an open system. The control of respiration by P_{CO_2} allows for the rapid destruction of bicarbonate when acid is added to the body, so that bicarbonate is better able to accept protons and functions as a stronger base than might be expected. When it takes up hydrogen ions, bicarbonate is converted to carbonic acid and disappears. The carbon dioxide formed from the carbonic acid is removed by the lungs, and only water is left. The amount of acid buffered is equal to the amount of bicarbonate which disappears. More than two-thirds of the bicarbonate initially present in the blood would be destroyed in bringing the pH from 7·4 to 7·0. The continual supply of carbon dioxide from cellular respiration, however, permits bicarbonate to be manufactured rapidly in almost unlimited quantities when cations are left unmatched by 'fixed' anions. Phosphate makes a far smaller contribution than bicarbonate despite its more favourable buffer ratio (4 : 1 at pH 7·4 compared with 20 : 1 for bicarbonate) because its concentration in the plasma is so much smaller.

The figure in Table 5 for the total buffer capacity shows that the blood of an ordinary adult (a little over 5 litres) could take up about 140 mmol of acid without a lethal fall in pH. 140 mmol may not seem a great deal, but it should be compared with the daily output of strongly acid end-products of metabolism. Since this is of the order of 50 to 80 mmol, the blood could take up the entire daily output of strong acid with a fall in pH of about 0·2 unit. In fact the blood is only required to hold acid products of metabolism whilst they are in transit from the sites of their production to the kidneys for excretion or to other tissues which assist in buffering them. The available buffering capacity of the blood should therefore be ample for ordinary purposes.

It is worth noting that although plasma is the fluid that is usually analysed, the erythrocytes contain a substantial part of the buffers in the blood. Hastings (1966) stated that whereas the plasma contains 28 mmol of bicarbonate and 16 mmol of other buffer anions (mostly protein and phosphate) per kg of water, the erythrocytes have 20 mmol of bicarbonate and 42 mmol of other buffer anions (mostly haemoglobin) per kg of water.

(ii) *Buffers in the Cells*

The nature and the capacity of the buffers in the cells are less completely known than those of the buffers in the plasma. But it is clear that the cells contain larger stores of buffer bases than does the blood. Hastings (1966) pointed out that the cells of the tissues, which make up two-thirds of the weight of the body, contain a smaller concentration of bicarbonate than the plasma (about 10 mmol/kg water), but have something of the order of 30 mmol of protein and 140 mmol of phosphoric esters per kg of water, which are also buffer anions. Despite some uncertainty in detail about their nature, there is evidence that the intracellular buffers share in the buffering of acids or alkalis which are added to the blood. Two examples will be sufficient to serve as illustrations. Swan, Pitts & Madisso (1955) infused hydrochloric acid into nephrectomized dogs, and found that only about one-fifth of the hydrogen ion was buffered in the blood and about one-quarter in the remainder of the extracellular fluids. About one-half seemed to have entered cells (in exchange for sodium and potassium which emerged into the extracellular fluids) and was presumably buffered by intracellular buffers. Singer et al. (1955) investigated the buffering of alkali by infusing hypertonic solutions of sodium bicarbonate intravenously into normal human subjects. About one-fifth of the alkali that remained in the body two hours after the infusions seemed to have been buffered in the cells. The participation of buffers throughout the body in the control of reaction was admirably reviewed in detail by Elkinton (1956); Winters & Dell (1965) have also tabulated valuable information on buffering outside the extracellular compartment.

Hydrogen ions which enter cells from the extracellular fluid tend to be replaced by metallic cations which move from the cells to the extracellular fluid, so that an exchange of ions takes place which to some extent supplements ordinary extracellular buffering. As stated on p. 94, hydrogen ions taken up by cells cannot remain free in the intracellular fluids. They are almost wholly combined with buffer anions, with the result that an equivalent number of negative charges disappear inside the cells, and cations must be released to maintain electrical neutrality. It is a remarkable thing that when this occurs cells seem able to retain their stores of potassium and part first with a substantial fraction of their far smaller intracellular stores of sodium (Pitts, 1953, 1963).

C. Exchanges of Ions with Mineral Constituents of the Skeleton

The mineral portion of the skeleton is believed to be present in the form of a vast number of tiny, flat, hexagonal crystallites, each about 50 nm across and 5–10 nm thick. This fine subdivision implies an enormously high ratio of superficial area to volume, which has been estimated as 100 square metres per gram of dry mineral substance. The total surface of the crystallites in an adult man's skeleton would be about 100 acres. Cations such as sodium, potassium, calcium and magnesium seem to exist in a loosely combined form at the surface of the crystallites, so that they can exchange with cations in the surrounding extracellular fluid. The skeleton contains about one-third of the sodium in the body, and the use of radioactive isotopes has shown that as much as one-third of the sodium in the skeleton may be exchanged with sodium in the extracellular fluid in the course of a day. A somewhat smaller proportion of the skeletal sodium seems to be exchangeable for hydrogen ions, and this enables the bones to supplement the buffering of acids which are added to the blood (Bergstrom, 1956; Nichols & Nichols, 1956).

The exchange of hydrogen ions with the mineral of the bones is not quite the same as ordinary buffering. When hydrogen ions are taken up by buffer bases, they disappear and are not replaced by other cations. Electrical neutrality is preserved by the loss of one negative charge by the buffer anion for each hydrogen ion taken up:

$$H^+ + Buf^- \rightarrow HBuf. \qquad (9)$$

When hydrogen ions are taken up by bone mineral, they are replaced in the plasma by other cations from the surface of the crystallites:

$$\begin{aligned} &H^+ + (Na^+\text{-, or } K^+\text{-, or } Ca^{2+}\text{-in-bone}) \\ &\quad \rightarrow (H^+\text{-in-bone}) + Na^+ \text{ (or other cation).} \qquad (23) \end{aligned}$$

This mechanism is not merely of academic interest; it is well known that prolonged acidosis may be associated with decalcification of the skeleton as the calcium ions removed from the bone mineral are excreted in the urine (Fourman, 1960).

Chapter 16

DISPOSAL OF NON-VOLATILE ACIDS BY THE KIDNEY

Strong acids which are absorbed or formed by metabolic reactions dissociate into hydrogen ions and the anions which are their weak conjugate bases. It is the hydrogen ions which tend to lower the pH of the body fluids, and in order to restore a normal reaction the kidneys require to excrete the hydrogen ions as well as the anions. The excertion of the anions usually presents little difficulty (unless the rate of glomerular filtration is grossly diminished) for they exist free in the plasma and enter the glomerular filtrate; all that is necessary is for the tubules to refrain from reabsorbing them. The acidosis which follows the absorption of hydrochloric acid or the metabolism of ammonium chloride cannot, however, be corrected by excreting the excess chloride ions together with, for example, the sodium and potassium which accompany them and balance their electrical charges in the plasma and in the glomerular filtrate. The excretion of chloride with sodium would leave the body no less acid, and it would remove a most important cation from the extracellular fluid.

The excretion of the hydrogen ions responsible for acidosis presents the kidneys with a problem, because, unlike the anions, these hydrogen ions no longer exist. They have nearly all disappeared by combining with buffer bases in the extracellular fluids and in the cells, or by exchanging with other cations from the bones, as described in the last section. Acidosis, as pointed out earlier, is marked by a reduction in the amount of buffer base in the plasma rather than by a fall in pH. Even if the pH were reduced to 7·0, the actual concentration of hydrogen ions would be only 0·0001 mmol/l, so that a normal renal plasma flow of about 1000L/day would present only 0·1 mmol/day to the kidneys to be excreted. Since most of the hydrogen ions derived from strong acids no longer exist free in the body, the blood cannot bring them to the kidneys to be excreted. The kidneys have no direct access to these hydrogen ions, and have to excrete them indirectly.

The kidneys in fact generate new hydrogen ions, equivalent in amount to those which need to be eliminated, and excrete these new hydrogen ions actively into the urine. They appear to be excreted in exchange for an equal number of sodium ions, which are reabsorbed from the glomerular filtrate and returned to the plasma. Although the manner in which this exchange of ions is brought about has not been elucidated in detail, it appears that one bicarbonate ion is also added to the plasma for each hydrogen ion which is excreted in the urine. The final results of the exchange are that the urine is made acid, the plasma is made alkaline by the bicarbonate added to it, and extracellular sodium is conserved. The last point is important, for about 50 to 80 mmol of acid has to be eliminated every day, and the total store of sodium in the extracellular fluid of an adult man is little more than 2,000 mmol. This would soon be exhausted if it were not for the kidneys' ability to replace sodium by a freshly made alternative cation to preserve electroneutrality in the urine.

The bicarbonate which is added to the plasma acts in three ways to combat acidosis and to restore both the normal reaction and the normal buffering capacity of the body fluids.

(1) It raises the concentration of bicarbonate in the plasma by replacing bicarbonate which had been destroyed by combining with hydrogen ions from acids stronger than carbonic.

(2) As the concentration of bicarbonate increases and the plasma becomes more alkaline, the reactions by which hydrogen ions had been buffered in the plasma and in the cells are reversed. Hydrogen ions are removed once more from the

conjugate acids of buffer bases, and the original buffer bases are regenerated. In general:

$$\begin{aligned} HBuf &\rightarrow H^+ + Buf^-; \\ H^+ + HCO_3^- &\rightarrow H_2CO_3; \\ \text{or, } HBuf + HCO_3^- &\rightarrow Buf^- + H_2CO_3. \end{aligned} \tag{24}$$

The equations for the regeneration of proteinate and phosphate buffers may be written:

$$HProt^{(n-1)-} + HCO_3^- \rightarrow Prot^{n-} + H_2CO_3, \tag{25}$$

and

$$H_2PO_4^- + HCO_3^- \rightarrow HPO_4^{2-} + H_2CO_3. \tag{26}$$

(3) The exchange of ions with the mineral parts of the skeleton is also reversed as the extracellular fluid becomes more alkaline:

$$\begin{aligned} &(H^+\text{-in-bone}) + Na^+ + HCO_3^- \\ &\quad \rightarrow (Na^+\text{-in-bone}) + H_2CO_3. \end{aligned} \tag{27}$$

Negative charges on bicarbonate ions disappear in reactions 24 to 26 but are replaced by an equal number of negative charges on the regenerated buffer anions. These negative charges are partly balanced by the positive charges of sodium ions reabsorbed along with the bicarbonate added to the plasma; partly made way for by the loss of negative charges on those anions such as chloride and sulphate which leave the plasma in the glomerular filtrate and are not reabsorbed by the renal tubules. The hydrogen ions of acids which were originally threatening the bland alkalinity of the plasma are finally combined with bicarbonate to form carbonic acid, and this breaks down to carbon dioxide and water. The lungs remove the carbon dioxide; the hydrogen ions remain harmlessly as hydrogen atoms in body water; and so it turns out that the actual hydrogen ions responsible for an acidosis are never excreted at all. The kidneys excrete new hydrogen ions into the urine, the lungs remove an equivalent amount of carbon dioxide, and the various buffer systems of the body are at last restored to their normal condition.

The renal mechanism for secreting acid must now be considered in more detail before its regulation to meet the body's changing needs can be discussed.

Chapter 17

SECRETION OF ACID BY THE RENAL TUBULAR EPITHELIUM

Human urine can be acidified to a pH a little below 4·6. The concentration of hydrogen ions is about 800 times greater in urine at pH 4·6 than in plasma at pH 7·4, and this implies that hydrogen ions are secreted against a large gradient of concentration. The hypothesis that the renal tubules make the urine acid by actively secreting hydrogen ions into it has indeed become generally accepted since Pitts and Alexander showed, in 1945, that no other current hypothesis could account for the rate at which dogs secreted acid after they had been made acidotic and given phosphate or creatinine to increase the amount of buffer in their urine. Creatinine, with pK′ 4·97, is a weak acid of which the conjugate base can be an effective urinary buffer (Pitts, 1968).

The hydrogen ions excreted by the kidney are usually assumed to arise from the dissociation of carbonic acid, formed from water and some of the carbon dioxide which is liberated during the respiration of the cells. The epithelial cells are rich in carbonic anhydrase, and the kidney's capacity to make the urine acid is drastically reduced if this enzyme is inactivated by inhibitors such as acetazoleamide (Diamox).

The reaction

$$H_2O + CO_2 \text{—(Carbonic anhydrase)}\rightarrow H_2CO_3 \rightarrow H^+ + HCO_3^-, \qquad (19)$$

may be regarded as an indirect means of obtaining hydrogen ions from water. An alternative view is that hydrogen ions are derived from water directly, and that carbon dioxide is used as a source of buffer to protect the cells from damage by the hydroxyl ions which are left behind. This mechanism was first proposed for the secretion of hydrogen ions by the gastric mucosa (Davies, 1951). Carbon dioxide might be employed to buffer hydroxyl ions by first forming carbonic acid, which would then immediately yield hydrogen ions to combine with the hydroxyl ions:

$$CO_2 + H_2O \xrightarrow{\text{(Carbonic anhydrase)}} H^+ + HCO_3^-; \quad (19)$$

$$H^+ + OH^- \rightarrow H_2O. \quad (5)$$

Alternatively, carbonic anhydrase might catalyse the reaction of carbon dioxide with hydroxyl ions to form bicarbonate in one step:

$$CO_2 + OH^- \xrightarrow{\text{(Carbonic anhydrase)}} HCO_3^-. \quad (28)$$

Either the indirect process providing hydrogen ions for excretion from water through carbonic acid, or the process depending upon hydrogen ions derived from water directly, would be impaired by the inhibition of carbonic anhydrase. The indirect process would be impaired because less carbonic acid would be available as a source of hydrogen ions. The direct process would be impaired because if the hydroxyl ions left behind were not buffered, they would reduce the concentration of hydrogen ions inside the cells. This would both reduce the supply of hydrogen ions to be secreted, and increase the gradient of concentration against which they had to be transported from the cells to the tubular lumen. The secretion of acid by either process should not, however, be stopped completely by inhibition of carbonic anhydrase, because the necessary reactions can proceed at an appreciable rate in the absence of the enzyme. The rate of secretion of H^+ ions has indeed been shown to increase appropriately at higher partial pressures of carbon dioxide during inhibition of carbonic anhydrase (Seldin et al., 1959).

Each of these two possible mechanisms is primarily concerned with the source of the hydrogen ions which are secreted into the urine. Both methods of supplying hydrogen ions have an important consequence in common; namely, that hydrogen ions and bicarbonate ions are produced in equal numbers. The essence of the secretory process is that it is so coupled with the reabsorption of sodium that, for every hydrogen ion secreted into the urine, one sodium ion is reabsorbed, and one bicarbonate ion is added to the plasma. This preserves electroneutrality in urine, in plasma, and in the tubular cell.

It is still not known just how the hydrogen ions are transported against a gradient of concentration. Pitts (1959) suggested that the gradient between the cells and the tubular lumen might be small enough (owing to a low pH of the intracellular fluid), and the difference of electrical potential across the membrane large enough, for an appreciable fraction of the hydrogen ions which enter the urine to be transported passively. The three processes, secretion of hydrogen ions into the urine, secretion of bicarbonate into the plasma, and reabsorption of sodium, are so closely linked that if any one of them was coupled to a metabolic source of energy, this primary active process could theoretically be made to drive the other two by suitable restrictions on the permeability of the tubular cells. (See discussion by Leaf, 1960.)

Recent measurements of transtubular differences of electrical potential have established that sodium ions are reabsorbed from all segments of the nephron against an electrical gradient, so that the reabsorption of sodium must be an active process, requiring energy, even when the concentrations are similar on opposite sides of the tubular epithelium. A close agreement between the short-circuit current measured across the proximal tubular epithelium and the amount of electricity carried across by sodium ions also points to reabsorption of sodium as the major active process in this segment (Giebisch & Windhager, 1964). It is of great interest that some of the most strongly acid urines produced by men were reported by Schwartz et al. (1955)

who made the astonishing observation that salt-depleted subjects who were actively reabsorbing sodium under the influence of salt-retaining steroids responded to an infusion of sodium sulphate and enough bicarbonate to make the plasma slightly more alkaline by a dramatic fall in urinary pH to as low as 4·0 associated with increased rates of excretion of potassium and ammonium. It appeared that H^+ and K^+ ions were being dragged out of the cells by a gradient of electrical potential which became sharply steeper when Na^+, which was being avidly reabsorbed was suddenly partnered in the tubular fluid by the non-penetrating anion, sulphate, instead of mainly by the more readily reabsorbable chloride. An increase in trans-tubular potential difference when sulphate or ferrocyanide was substituted for chloride was actually demonstrated by micro-puncture techniques in analogous experiments on rats by Clapp, Rector & Seldin (1962). But the largest recorded potential differences were no greater than 60 to 70 millivolts, and so could not account for a concentration ratio for H^+ ions (tubular fluid/plasma) much greater than 10 to 1. It would need potential differences of over 180 mV (corresponding to a ratio of 1000 : 1) to account for the intense acidification observed by Schwartz et al. Hence although a substantial amount of the hydrogen ion added to the tubular fluid may move passively in response to a potential gradient set up by the active reabsorption of sodium, some kind of 'booster pump' for hydrogen ions is needed to complete the process of secretion into highly acid urines.

The detailed explanation of how hydrogen ions are secreted, either independently or in exchange for sodium must wait upon future discoveries. From the narrower point of view of the kidney's share in regulating the reaction of the body fluids, the most important feature of the exchange mechanism is the simple quantitative relation that one bicarbonate ion is added to the plasma for each hydrogen ion secreted into the urine.

Chapter 18

SITE OF RENAL TUBULAR SECRETION OF HYDROGEN IONS

Despite a certain amount of evidence to the contrary (Smith, 1951, 1956), it used to be taught that the urine was acidified in the distal tubule; and this was supposed to be the only site where hydrogen ions were actively secreted in exchange for sodium. Pitts & Lotspeich had, however, pointed out in 1946 that if any bicarbonate from the glomerular filtrate reached a part of the tubule into which hydrogen ions were being secreted, hydrogen ions would combine with the bicarbonate,

$$H^+ + HCO_3^- \rightarrow H_2CO_3,$$

and the carbonic acid formed would break down to water and carbon dioxide, most of which would diffuse across the tubular epithelium into the interstitial fluid and so into the blood. Since, however, a bicarbonate ion from the tubular epithelium would be added to the plasma for each hydrogen ion which was secreted into the urine, a number of bicarbonate ions would disappear from the lumen of the tubule, and an equal number would appear in the plasma. The final result would be tantamount to a reabsorption of bicarbonate, although the bicarbonate ions which were in the lumen would have been destroyed, and the bicarbonate ions which appeared in the plasma would be new ones which had been produced in the epithelial cells. Pitts & Lotspeich (1946) suggested that any bicarbonate which reached the distal tubules was 'reabsorbed' in this manner. Berliner later found that when carbonic anhydrase was inhibited, the kidneys did not merely cease to make the urine acid; the urine actually became alkaline, and contained up to half as much bicarbonate as there had been in the glomerular filtrate (Berliner & Orloff, 1956). It therefore appeared that the removal of at least half of the bicarbonate from the tubular lumen depended upon the secretion of hydrogen ions. Since it was unlikely that half the filtered bicarbonate would be left over to enter the distal tubules, it also appeared that carbonic anhydrase was involved in the 'reabsorption' of bicarbonate from the proximal tubule; and this in turn implied that the secretion of hydrogen ions occurred in the proximal as well as in the distal tubules. This was not necessarily inconsistent with reports that the urine first became acid in the distal tubule, because the addition of hydrogen ions would not be expected to make the tubular fluid acid until bicarbonate had been removed, and this might not ordinarily occur before the distal tubule was reached.

It has now been confirmed directly that the secretion of hydrogen ions is not confined to the distal tubule. Gottschalk et al. (1960) punctured the nephrons of rats at various levels and demonstrated that acidification could occur half-way along the proximal tubules, especially if the rats had been given ammonium chloride to create an acidosis. Further acidification occurred in the distal tubules, and the pH of urine from the ureter was still lower, suggesting that hydrogen ions were also secreted by the collecting ducts. This had been shown directly in hamsters by Ullrich et al. (1958a), who had cannulated the collecting ducts with fine polyethylene catheters, and found that the pH was lower in samples of fluid which had travelled farther along the collecting ducts. Although the secretion of hydrogen ions thus appears to occur in most parts of the nephron, investigators who used the recently developed 'Stop-flow' technique found in dogs that secretion was particularly intense at a distal site where sodium was avidly reabsorbed (Malvin et al., 1958; Pitts et al., 1958).

Chapter 19

REABSORPTION OF BICARBONATE

Bicarbonate is paramount among the body's buffer bases for three reasons. It is present in greatest concentration; it can be used to regenerate other buffers and to reverse the ionic exchanges which supplement buffering; and it can be manufactured in almost unlimited quantities from water and carbon dioxide. As ATP seems to function as a general currency for biological energy, so bicarbonate is a common currency for paying off all kinds of buffer debts. Since we lose between 4000 and 5000 mmol of bicarbonate from the plasma every day in the glomerular filtrate, the recovery of this valuable buffer base must be one of the first and most urgent tasks of the renal tubules.

The literature on the reabsorption of bicarbonate has become too complicated to review here in detail. The principal reasons for concluding that a substantial portion of filtered bicarbonate is 'reabsorbed' as a consequence of the tubular secretion of hydrogen ions may be very briefly summarized: 1. Pitts & Lotspeich (1946) demonstrated an upper limit to the rate at which bicarbonate was removed from the glomerular filtrate when its concentration in the plasma was raised. 2. The rate at which bicarbonate can be 'reabsorbed' is depressed by the inhibition of carbonic anhydrase (Berliner & Orloff, 1956). 3. The rate at which bicarbonate can be 'reabsorbed' from the glomerular filtrate varies roughly in proportion to the partial pressure of carbon dioxide in the blood (Brazeau & Gilman, 1953; Dorman et al., 1954). Goodman & Fuisz (1964) were able to show in dogs that the capacity of the left kidney to reabsorb bicarbonate was increased by increasing P_{CO_2} in the blood supplied to that kidney alone. Moreover, the capacity to reabsorb bicarbonate still varied with the carbon dioxide tension of the blood when carbonic anhydrase was inhibited (Seldin et al., 1959). The above observations suggest that bicarbonate is 'reabsorbed' by an indirect process which depends upon the production of carbonic acid.

4. When the urine is alkaline and contains bicarbonate, its P_{CO_2} has been found to exceed that of the plasma (Pitts & Lotspeich, 1946); but the elevated P_{CO_2} of alkaline urine can be reduced by infusing carbonic anhydrase intravenously so that the enzyme appears in the tubular fluid (Ochwadt & Pitts, 1956). When the urine contains bicarbonate, it may be presumed that carbonic acid formed by the addition of hydrogen ions to bicarbonate in the collecting ducts is carried into the lower urinary tract where, in the absence of carbonic anhydrase, it breaks down rather slowly and releases carbon dioxide in a place from which it can less easily diffuse back to the blood. (The bulk of fluid is far greater in relation to surface area in the lower urinary tract than in the nephron.)

More direct evidence that bicarbonate is 'reabsorbed' from the proximal tubule by decomposition with hydrogen ions was provided by the micropuncture method (Windhager, 1968). Rector, Clapp & Seldin (1965) argued that if bicarbonate was 'reabsorbed' in this way, there ought to be a 'disequilibrium pH' in the sense that until the H_2CO_3 formed had decomposed into CO_2 and water the tubular fluid should be more acid than it would become after reaching equilibrium with the prevailing P_{CO_2}. They used a glass micro-electrode to measure intratubular pH and a quinhydrone electrode to measure intratubular bicarbonate in rats; and they controlled the partial pressure of carbon dioxide. There was a disequilibrium pH of about 0·85 unit in the distal convoluted tubule, where it could be abolished by infusing carbonic anhydrase. It also disappears during acidosis when bicarbonate does not reach the distal tubule (Giebisch & Malnic, 1970). Normally there was no disequilibrium pH in the proximal tubule, presumably because carbonic anhydrase in the luminal border of the epithelium enabled equilibrium to be attained as quickly as the glass electrode could respond and measure the pH. A disequilibrium pH like that in the distal tubule appeared in the proximal tubule also if carbonic anhydrase was inhibited. Hence, it appeared that hydrogen ions are secreted into the proximal tubule and decompose bicarbonate. By preventing the setting up of a disequilibrium pH the luminal carbonic anhydrase must reduce the concentration ratio against which hydrogen ions have to be secreted by a factor of nearly 10.

It is not necessary for the whole of the filtered bicarbonate to be 'reabsorbed' in consequence of the tubular secretion of hydrogen ions as described above. A major fraction of the filtered *sodium* is probably reabsorbed by some kind of 'sodium pump', not in exchange for hydrogen ions, but accompanied by anions to preserve electrical neutrality. Most of these accompanying anions are almost certainly chloride; but some of the bicarbonate of the glomerular filtrate could, in principle, be reabsorbed passively in this way. Rector et al. (1965) concluded that most, if not all, of the filtered bicarbonate is, in fact, 'reabsorbed' by reacting with hydrogen ions secreted into the tubules. Maren (1969), however, considered that at least half the bicarbonate must be reabsorbed in some other way.

The significance of the indirect mechanism for 'reabsorbing' bicarbonate may depend upon the fact that the bicarbonate ion is larger than the chloride and even than the hydrated sodium ion (Ito et al., 1962). The luminal border of the tubular epithelium seems to have a low permeability to bicarbonate (Clapp et al., 1962; Malnic & Aires, 1970), which is therefore likely to be a nonpenetrating anion that cannot easily accompany reabsorbed sodium across the luminal cell membrane as chloride does. Yet the early micropuncture work of Walker et al. (1941) appeared to show that in rats bicarbonate had almost disappeared by the end of the proximal tubule, and was therefore being reabsorbed in preference to the presumably more permeant chloride. This requires some explanation, and the indirect mechanism offers a means of replacing bicarbonate from the tubular fluid by bicarbonate in the plasma without any bicarbonate ions actually having to cross the luminal cell membrane. The peritubular border of the epithelium is probably freely permeable (Wick & Fromter, 1967), so that bicarbonate formed in the cells can readily pass out into the plasma.

Chapter 20

CONSEQUENCES OF THE SECRETION OF HYDROGEN IONS INTO THE TUBULAR URINE

We can now consider the fate of the hydrogen ions which enter the tubular fluid in exchange for sodium, and the forms in which they may be excreted in the urine. Three important consequences which follow the addition of hydrogen ions to the fluid passing along the renal tubules may be described most conveniently as though they occurred in distinct segments. In fact, the three processes, which are depicted in Fig. 2, must overlap to some extent; and the portions of the nephron in which they take place may differ according to prevailing conditions.

(1) 'Reabsorption' of filtered bicarbonate. So long as bicarbonate remains in the tubule, it takes up any hydrogen ions which are secreted, and is itself destroyed by conversion to carbonic acid which breaks down to carbon dioxide and water. An equal amount of bicarbonate is added to the plasma by the mechanism which provides hydrogen ions to be secreted. The hydrogen ions themselves end up in water, which may remain in the urine, or diffuse out from the tubule through the interstitial fluid to the renal venous blood.

This first phase of the secretion of hydrogen ions in exchange for sodium completes the 'reabsorption' of bicarbonate from the glomerular filtrate. Normally, in man, about 3 mmol of bicarbonate enters the glomerular filtrate every minute, but little or none appears in the urine, which is somewhat acid on ordinary mixed diets. For filtered bicarbonate to be 'reabsorbed' at this rate, carbon dioxide formed in the tubular lumen by dehydration of carbonic acid must probably enter

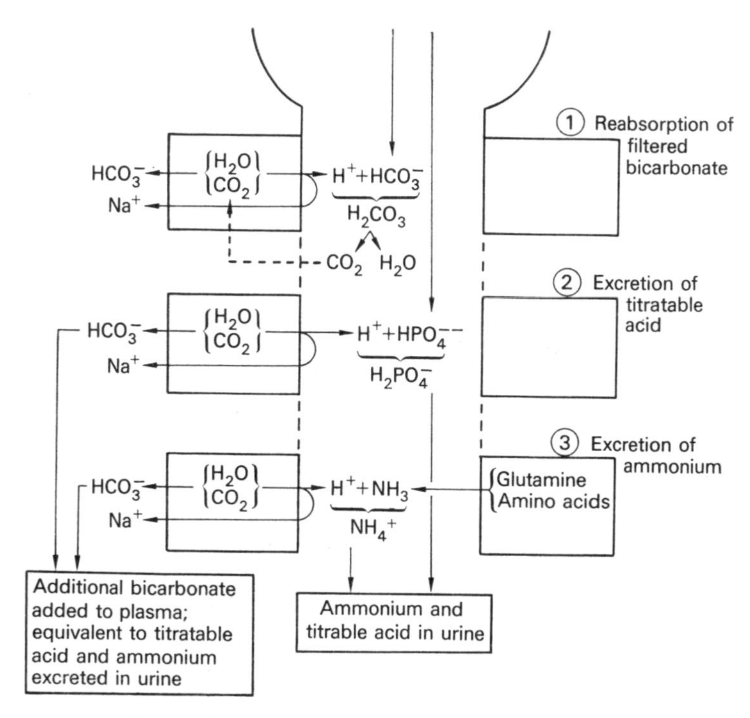

FIG. 2. Three phases of renal tubular ion exchange.

the cells and help to supply hydrogen ions for transfer into the tubular fluid and bicarbonate to pass into the plasma. The kidneys take about 10% of a man's resting consumption of oxygen, i.e., about 24 ml, or 1 mmol, per minute. Even if all this were converted to carbon dioxide, local cellular respiration could provide no more than 1 mmol/min of hydrogen and bicarbonate ions. Hence most of the carbon dioxide that is used to provide hydrogen ions must come from the blood or the tubular fluid. By hastening the dehydration of carbonic acid formed by hydrogen ions reacting with filtered bicarbonate, the carbonic anhydrase in the luminal border of the proximal tubule should increase $P\text{CO}_2$ locally and promote diffusion of carbon dioxide from the lumen into the cells.

If the rate at which bicarbonate appears in the glomerular filtrate is increased, or if the secretion of hydrogen ions is depressed, as by a low partial pressure of carbon dioxide or by the action of an inhibitor of carbonic anhydrase, the point at which bicarbonate has all disappeared and the urine becomes acid moves down the tubule. If there is more bicarbonate than the total amount of hydrogen ion secreted, the urine never becomes acid, and the excess bicarbonate is excreted in an alkaline urine. The common finding of a high partial pressure of carbon dioxide in such an alkaline urine has already been mentioned as evidence in favour of the view that bicarobnate is removed, at least from the distal nephron, by decomposition rather than by reabsorption as such.

It should be noted that the complete 'reabsorption' of filtered bicarbonate does not make the urine acid, and does nothing in itself to correct an acidosis. It merely ensures that as much bicarbonate is returned to the plasma as had left it in the glomerular filtrate. Correction of an acidosis requires the addition to the plasma of more bicarbonate than is lost by glomerular filtration.

(2) Excretion of titratable acid. Hydrogen ions which are added to the tubular urine after bicarbonate has been removed are taken up by other buffers, and convert the buffer bases to their conjugate acids. Thus, for example, secondary phosphate is converted to primary phosphate:

$$HPO_4^{2-} + H^+ \rightarrow H_2PO_4^-. \qquad (7)$$

Hydrogen ions secreted by the tubular epithelium thus appear in the urine, largely unionized, in molecules of the conjugate acids of buffer bases which were present in the glomerular filtrate. These conjugate acids constitute the 'titratable acid' of the urine. The amount of titratable acid which has been excreted is discovered by titrating the urine back to the pH of the plasma with standard alkali. During the titration the hydrogen ions are removed once more to combine with hydroxyl ions from the alkali, so that the process by which titratable acid was formed in the kidney is reversed *in vitro.*

This second phase of the secretion of hydrogen ions gives the urine an acid reaction. The amount of titratable acid produced depends upon the capacity of the urinary buffers and the pH finally reached. At the lower limit, around pH 4·6, phosphate is virtually completely converted to $H_2PO_4^-$, and two-thirds of the creatinine is present in the acid form. In healthy people on ordinary mixed diets the urine contains some 20 to 30 mmol of titratable acid per day. In diabetic ketosis, when β-hydroxybutyric acid (pK′ 4·7) and acetoacetic acid (pK′ 3·6) contribute to the titratable acidity, the total may rise to as much as 250 mmol per day (Pitts, 1968).

The excretion of titratable acid in the urine assists in removing acid from the body, and in restoring reserves of buffer base in the plasma and elsewhere. For every hydrogen ion excreted as titratable acid in the urine, an additional bicarbonate ion is added to the plasma over and above those bicarbonate ions which merely replace bicarbonate lost in glomerular filtration.

(3) Excretion of ammonium. Ammonia is formed in the tubular epithelium, partly from glutamine through the action of a glutaminase present in the cells, and partly by the oxidative deamination of amino acids. As hydrogen ions are added to the tubular urine to make it acid, ammonia which diffuses out from the cells combines with hydrogen ions in the lumen and forms ammonium:

$$NH_3 + H^+ \rightarrow NH_4^+. \qquad (29)$$

This process was thought to take place principally in the distal tubule, but Ullrich et al. (1958b) have shown that it can occur in the collecting ducts. The hydrogen ions secreted in this third phase of their exchange for sodium appear in the urine as ammonium. Ammonium is a newly formed, and consequently expendable, cation which can replace the sodium which is reabsorbed. It is also a weak acid (Table 1), for it yields up its proton and reforms ammonia in the presence of a base stronger than ammonia. The peculiar physiological significance of the formation of ammonium in the kidney is that it permits the secretion of hydrogen ions in exchange for sodium to continue without further lowering the pH of the urine. A great deal more hydrogen ions can therefore be secreted (and a corresponding amount of additional bicarbonate can be added to the plasma) after the urinary buffers have been brought to the lower limit of urinary pH.

Under ordinary conditions 30 to 50 mmol of the hydrogen ion secreted each day to acidify the urine is excreted as ammonium, but this can be increased to as much as 500 mmol per day in severe sustained acidosis. It takes several days for the largest rates to be attained, but the mechanism of this gradual, delayed increase is not yet clear. It is not necessarily accompanied by a progressive fall in urinary pH, and it cannot be satisfactorily explained by an adaptive increase in activity of the enzymes involved. It might, however, be related to an augmented secretion of hydrogen ions. A tubular fluid that was initially more strongly acid would need more ammonia to bring it to a given final pH than a tubular fluid that started less acid. Urine derived from an initially more acid tubular fluid would therefore have to contain more ammonium than urine of the same final pH derived from a tubular fluid that was less acid before taking up ammonia. In so far as the secretion of hydrogen ion is secondary to the active reabsorption of sodium, it might come under the control of the adrenal cortex. Adrenalectomized animals and patients with Addison's disease are notably poor at acidifying their urine and excreting ammonium. The slow adaptive increase in excretion of ammonium might be explained by more intense secretion of hydrogen ion stimulated by aldosterone mobilized under the threat of depletion of the body's stores of sodium during a prolonged acidosis.

Pitts and his co-workers have carried out a remarkably thorough study of the sources of urinary ammonia (Pitts, 1968, 1971). About 40% comes from the amide N and 20% from the amino N of glutamine. A further 5% comes from alanine, 3% from glycine and 2% from glutamate. The remaining 30% comes from the arterial blood, but much of this probably comes ultimately from the kidneys. Ammonia formed in the epithelial cells leaves the kidneys in venous blood and urine in proportions which depend upon their respective pHs. If these are similar most ammonia is removed by the far greater volume of venous blood, and under these conditions the rate of excretion varies with the flow of urine (Macknight et al., 1962). When the urine is being acidified, however, the rate of production of ammonia increases, an increasing proportion of the ammonia formed is removed in the acid urine, and the rate of excretion is determined largely by the pH of the urine (See also Richterich, 1962).

In so far as a bicarbonate ion is added to the plasma for each hydrogen ion secreted into the tubular urine, the total effect of the three phases of hydrogen ion secretion illustrated in Fig. 2 is to add to the plasma as much bicarbonate as was removed by glomerular filtration (Phase 1), and then as much more bicarbonate as the total amount of acid excreted in the urine. The total amount of acid excreted (the sum of titratable acid and the much weaker acid, ammonium) is therefore a measure of the kidney's daily share in the task of eliminating acid from the body and in preserving the blandly alkaline reaction of the plasma. The normal quantity of acid excreted in the urine (20 to 30 mmol of titratable acid and 30 to 50 mmol of ammonium) corresponds to the 50 to 80 mmol of strong acid produced by metabolism each day.

It is interesting to notice how small a fraction of the quantity of hydrogen ions secreted by the renal tubular epithelium actually reaches the urine—even in the mainly masked forms of titratable acid and ammonium. If the concentration of bicarbonate in the plasma were to be 25 mmol/l and the rate of glomerular filtration 180 l/day, 4,500 mmol of hydrogen ion would be required for the 'reabsorption' of bicarbonate on the assumption that no unchanged bicarbonate is reabsorbed directly. A normal daily excretion of 50 to 80 mmol of urinary titratable acid and ammonium would require less than 2% of the total amount of hydrogen ion secreted by the renal tubular cells. Even if a substantial fraction of filtered bicarbonate were reabsorbed directly by a process unconnected with hydrogen ion secretion, urinary acid excretion would require no more than a small fraction of the hydrogen ions secreted.

It is also worthy of note that the bulk of the hydrogen ions secreted into the tubule, that part which is not destined for excretion in the urine but will be used up in the 'reabsorption' of filtered bicarbonate, does not appreciably acidify the tubular fluid—even temporarily—where there is luminal carbonic anhydrase to avoid the setting up an a disequilibrium pH. The major part of the hydrogen ion is therefore secreted against no more than a minimal gradient of concentration, and it may well be transferred passively in response to gradients established by the sodium pumps responsible for reabsorption of sodium. It is quite otherwise for at least part of the smaller fraction of secreted hydrogen ion that is destined for excretion in acid urine. This fraction may have to be transferred against substantial gradients, and it is for this that the 'booster pump' mentioned on page 68 may be required. The absence of luminal carbonic anhydrase from the distal tubule means that hydrogen ions have to be transferred against a steeper gradient. The greater acidity of the tubular fluid (indicated by the disequilibrium pH) must, however, promote the diffusion of ammonia from the epithelial cells into the urine.

Chapter 21

EXCRETION OF ACID AND ALKALINE URINES

The total amount of hydrogen ion *secreted* by the tubular epithelium is greater than the total amount *excreted* as urinary acid (titratable acid and ammonium). The amount of hydrogen ion which is left over to make the urine acid depends upon how much has been used up in 'reabsorbing' bicarbonate. If the amount of bicarbonate filtered is more than equivalent to all the hydrogen ion secreted, no hydrogen ions get through to the final urine; instead, bicarbonate ions are left over, and the urine is alkaline. If insufficient bicarbonate is filtered to take up all the hydrogen ions which the tubules secrete, then there are hydrogen ions left over, and the urine is acid. The amount of acid or alkali excreted in the urine is therefore mainly determined by the difference between the rate at which bicarbonate is filtered and the rate at which the tubules secrete hydrogen ions.

The rate at which bicarbonate appears in the glomerular filtrate is proportional to the concentration of bicarbonate in the plasma and to the rate of glomerular filtration. Normally in man the rate of glomerular filtration is rather constant at about 125 ml/min, so that the rate of filtration of bicarbonate is simply proportional to the concentration of bicarbonate in the plasma. If this has the usual value of about 24 mmol per litre of plasma, then about 3 mmol of bicarbonate is filtered into the tubules each minute.

The rate at which hydrogen ions are secreted is determined principally by the partial pressure of carbon dioxide in the plasma, to which it is roughly proportional (References, page 71). The partial pressure of carbon dioxide might determine the rate of secretion by affecting the pH inside the cells as well as by controlling the supply of carbonic acid to the secretory mechanism. The reaction of the intracellular fluid is also affected by the amount of potassium in the cells (see later, page 94). A deficiency of potassium leaves the cells unduly acid, and the kidney tends to secrete an inappropriately acid urine. Meanwhile the addition of excessive amounts of bicarbonate to the plasma may lead to the development of an extracellular alkalosis.

Chapter 22

EFFECT OF CONCENTRATION OF BICARBONATE IN PLASMA

Of the various factors which influence the amount of acid excreted in the urine, the rate at which bicarbonate passes into the glomerular filtrate seems to be the most important. The effect of this factor in isolation is illustrated by the hypothetical, but not unreasonable, figures in Table 6. It is assumed for simplicity that the rate of glomerular filtration is constant at 125 ml/min, that all the filtered bicarbonate is 'reabsorbed' by reacting with hydrogen ions, and that no respiratory compensation occurs to alter the partial pressure of carbon dioxide, so that the rate of secretion of hydrogen ions is constant.

The third row of figures in Table 6 represents normal conditions. The upper two rows show the effect of increasing the concentration of bicarbonate and the lower two rows the effect of reducing the concentration of bicarbonate in the plasma. The concentration of bicarbonate determines whether the urine shall be alkaline or acid, and also the rate at which excess alkali or acid would be excreted in it if no other factors besides the concentration of bicarbonate in the plasma affected the result. It will be seen that alterations in the concentration of bicarbonate in the plasma lead automatically to changes in renal function which are appropriate for the correction of acidosis or of alkalosis. In metabolic alkalosis the concentration of bicarbonate in the plasma is raised, so that more bicarbonate is filtered and less hydrogen ions reach the urine. If more bicarbonate reaches the tubules that can be decomposed by all the hydrogen ions which are secreted, the urine contains the excess bicarbonate and becomes alkaline. In acidosis the concentration of bicarbonate in the plasma is reduced.

TABLE 6. Renal compensation for acidosis and alkalosis (respiratory compensation neglected: GFR 125 ml/min)

$[HCO_3^-]$ *in plasma*	HCO_3^- *filtered*	H^+ *secreted* = HCO_3^- *added to plasma.*	*URINE*		
			Reaction	*Content*	*mmol /min*
mmol /l	*mmol /min*	*mmol /min*			
48	6·0	3·05	ALKALINE	Bicarbonate	2·95
36	4·5	3·05	ALKALINE	Bicarbonate	1·45
24	3·0	3·05	ACID	Titr. acid $+NH_4^+$	0·05
18	2·25	3·05	ACID	Titr. acid $+NH_4^+$	0·80
12	1·5	3·05	ACID	Titr. acid $+NH_4^+$	1·55

This reduces the amount of bicarbonate filtered and leaves more hydrogen ions to be excreted as acid in the urine. Not so much more hydrogen ion as the table indicates is excreted in the urine, however. This is because the rate of secretion of hydrogen ions depends upon the partial pressure of carbon dioxide in the blood supplied to the kidneys. In metabolic acidosis the arterial P_{CO_2} is lowered by the increased pulmonary ventilation which occurs when the peripheral chemoreceptors respond to the falling pH of the plasma, so that the rate of tubular secretion of hydrogen ions is reduced during metabolic acidosis.

Chapter 23

INFLUENCE OF RESPIRATORY COMPENSATION

The excretion of 0·05 mmol of total acid per minute (shown in the third row of Table 6) corresponds to 72 mmol per day, which is the sort of rate that might be expected under normal conditions. (The rate of *secretion* of hydrogen ions was in fact assumed to be 3·05 mmol/min in order that the calculated rate of *excretion* of acid should fall in the normal range.) 0·80 mmol/min (fourth row of Table 6) amounts to 1152 mmol/day, which is a far greater output of titratable acid and ammonium than would be expected in moderate acidosis. This is because the respiratory response and its effect upon the kidney were ignored. In an actual acidosis, respiratory compensation would reduce the partial pressure of carbon dioxide in the blood, and the rate of secretion of hydrogen ions by the kidney would be less than 3·05 mmol/min. If, for example, the partial pressure of carbon dioxide was reduced to 30 mmHg and the rate of secretion of hydrogen ions to 2·55 mmol/min. There would still be hydrogen ions left over after all the filtered bicarbonate had been dealt with. The urine would therefore be acid, and the rate of excretion of total urinary acid (titratable acid plus ammonium) would be 2·55 — 2·25 = 0·30 mmol/min, or 432 mmol/day, a more plausible figure. Under these conditions the kidneys would be adding an extra 432 mmol of bicarbonate to the plasma each day over and above the amount which had been lost by glomerular filtration.

The total rate of secretion of hydrogen ions under these conditions amounts to 2·55 mmol/min, or 3672 mmol/day, which may be compared with 3·05 mmol/min or 4392 mmol/day under normal conditions (third row of Table 6). During the acidosis, however, 432 of the 3672 mmol of hydrogen ions, or 12%, would be employed in the formation of urinary titratable acid and ammonium, compared with 72 of 4392 mmol or 1·65% under normal conditions. The paradox that the rate of secretion of hydrogen ions may be reduced when the kidneys are excreting acid more rapidly in response to metabolic acidosis was pointed out by Gilman (1958). The amount of hydrogen ion which the kidneys excrete against a gradient from plasma into acid urine is, however, greater during the correction of acidosis, and more energy will be required for the process, even though the total turnover of hydrogen ions may be less.

Respiratory compensation of the change in reaction produced by an excess of acid in the body slows the secretion of hydrogen ions by the renal tubules so that it takes longer to get rid of the unwanted acid. Since at the same time the reduction in the partial pressure of carbon dioxide largely prevents a fall in pH while the acidosis persists, the longer time required to excrete all the acid may be a small price to pay. The greater duration of the acidosis causes relatively little inconvenience so long as the accompanying acidaemia is compensated.

We might expect that the alkalaemia which accompanies a metabolic alkalosis would be similarly compensated for by retention of carbon dioxide in the body as a result of a diminished rate of pulmonary ventilation. But, as Moran Campbell (1968) pointed out, the compensatory increase in $P\text{CO}_2$ is often small and may be absent, possibly because an increasing $P\text{CO}_2$ makes the cells and the cerobrospinal fluid more acid and this nullifies the inhibition of breathing that should occur if the peripheral chemoreceptors were acting alone. To the extent that respiration is inhibited, the higher partial pressure of carbon dioxide increases the rate at which the renal tubules secrete hydrogen ions, so that more of the filtered bicarbonate is decomposed. The urine therefore contains less bicarbonate than the first two rows of Table 6 suggest, and it takes longer to eliminate the excess alkali from the body. But the longer duration of the alkalosis is of less consequence when the associated alkalaemia is partly compensated.

Robin & Bromberg (1959) raised an interesting point in connection with the respiratory compensation for disturbances of reaction. Cell membranes seem to be freely permeable to carbon dioxide and to unionized carbonic acid, but far less permeable to bicarbonate ions and to hydrogen ions. Alterations in the partial pressure of carbon dioxide thus alter the concentration of carbonic acid throughout the fluids of the body in a few minutes, but the concentration of bicarbonate in the cells and in transcellular fluids such as the cerebrospinal fluid may take several hours to catch up with a change in the concentration of bicarbonate in the plasma. Thus when acid is added to the blood, the concentration of bicarbonate falls rapidly in the plasma but not in the cells and the cerebrospinal fluid. Meanwhile the increased ventilation, which results from the increasingly acid reaction of the plasma, quickly lowers the partial pressure of carbon dioxide and the concentration of carbonic acid throughout the body. The result is that, at least for a time, the cerebrospinal fluid and the contents of the cells become more alkaline, while the plasma is more acid than normal. The reactions of the plasma and of the cerebrospinal fluid are also altered in opposite directions during acute metabolic alkalosis if alkalinity of the plasma reduces ventilation, increases $P\text{CO}_2$ and makes the cerebrospinal fluid more acid. The administration of sodium bicarbonate has been found to make the blood more alkaline and the cerebrospinal fluid more acid.

It has already been observed that the pH of the cerebrospinal fluid appears to control pulmonary ventilation through intracranial chemoreceptors, and that this tends to stabilize the reaction of the fluids that bathe the brain. If the cerebrospinal fluid, so far as concerns the direction of changes in its reaction, is a valid model for the intracellular fluids of the body, the intracranial chemoreceptors may also play an important part in controlling the reaction of the intracellular fluids. We may picture the intracranial chemoreceptors standing guard over the reaction of the intracellular fluids and the brain whilst the peripheral chemoreceptors guard primarily the blood and the more accessible extracellular fluids. In fact the response of the peripheral receptors to hypoxia is probably more important than their response to $P\text{CO}_2$ or pH. They may be regarded as primarily concerned with controlling respiration so as to safeguard the supply of oxygen to the brain, although, in face of gaseous disturbances of reaction (which affect cells, cerebrospinal fluid and plasma in the same direction) they reinforce the action of the central chemoreceptors and help to preserve the reaction of all the body fluids. During metabolic disturbances conflicts may arise and the action of the peripheral receptors to stabilize the pH of the extracellular fluids may be over-ridden by the intracranial receptors in the interest of the cerebrospinal fluid, the brain, and the cells generally.

Chapter 24

RENAL COMPENSATION OF RESPIRATORY DISTURBANCES

The effect of the partial pressure of carbon dioxide upon the renal secretion of hydrogen ions provides an automatic compensation for disturbances of respiratory origin.

In respiratory acidaemia the partial pressure of carbon dioxide is high. The concentration of bicarbonate in the plasma is also somewhat increased, and more bicarbonate is filtered. If it were not for the increased rate of secretion of hydrogen ions which accompanies the higher partial pressure of carbon dioxide in the plasma, the faster filtration of bicarbonate would lead to the excretion of an alkaline urine. In fact the secretion of hydrogen ions usually increases more than the filtration of bicarbonate, and the urine becomes acid. Meanwhile in the plasma, the concentration of bicarbonate, already raised by interaction with the other buffer systems (page 20), is increased still further by the action of the kidneys (Fig. 1), and the buffer ratio shifted towards normal.

In respiratory alkalaemia, the concentration of bicarbonate in the plasma is low and the rate of filtration of bicarbonate is diminished in proportion. This alone would tend to leave more hydrogen ions to be excreted as titratable acid and ammonium. But the low partial pressure of carbon dioxide in the plasma reduces the total secretion of hydrogen ions so much that the urine becomes alkaline and contains bicarbonate in spite of the low concentration of bicarbonate in the plasma. This renal response, paradoxical at first sight, is easily explained on the hypothesis that filtered bicarbonate is 'reabsorbed' by decomposition and replacement as has been described earlier.

It has long been known (Peters & Van Slyke, 1931) that the

urine may contain 'acetone bodies' after periods of overbreathing. Lotspeich (1959) described how, when the urine is alkaline, increasing quantities of many anions, which are the conjugate bases of acids occurring as intermediates in the metabolic cycles of the renal cells, begin to be excreted. The kidney seems to be using their negative charges to balance the positive charges on metallic cations excreted in the alkaline urine, much as it has been supposed to manufacture ammonium, a new cation, to excrete with unwanted anions when sodium is being conserved. The apparently purposive response may however be no more than a consequence of the tendency of the unionized acids to diffuse from the cells to an alkaline urine as ammonia does to an acid urine.

Cell membranes generally seem more permeable to uncharged species than to the corresponding ions. Many weak acids and bases become unevenly distributed between the two sides of a membrane across which a difference of pH is maintained. The uncharged molecular species exhibits a net diffusion across the membrane towards the side on which the prevailing reaction favours ionization and keeps the concentration of the unionized form low. The total concentration (of ionized and unionized forms together) thus tends to become greater on this side, and the substance is sometimes said to be 'trapped' because the ionized form cannot readily diffuse back down its concentration gradient on account of the comparative impermeability of the membrane. The substance can then be removed from the system in a stream of flowing solution. This principle of 'non-ionic diffusion' was clearly set out by Milne, Scribner & Crawford (1958). It has since been applied extensively to account for the excretion and back-diffusion of a number of weak acids and bases (Weiner & Mudge, 1964), one of the most important of which is ammonia (Pitts, 1964). This same mechanism may operate to some extent between the cells and the slightly alkaline extracellular fluids. Concentrations of lactic and pyruvic acid in the blood increase during acute respiratory alkalaemia, and this has the appearance of a compensatory response to mitigate the increase in pH. Campbell (1968), however, suggested that the increase in concentration of organic acids might be no more than an incidental consequence of the greater alkalinity of the plasma.

Chapter 25

POTASSIUM AND THE RENAL SECRETION OF ACID

It is an oversimplification to regard hydrogen ions as the only cations which the renal tubular cells can excrete in exchange for sodium. Potassium and hydrogen ions have been described as competing for a common pathway of excretion because of reciprocal relations between the rates of their appearance in the urine (Berliner & Orloff, 1956). A looser and less direct coupling may now be more likely than the sharing of a specific common mechanism or secretory pathway (Giebisch et al., 1971); but intracellular pH appears to be a major determinant of the excretion of potassium, and it remains true that sodium appears to be reabsorbed in exchange for either hydrogen or potassium according to the relative availability of these two cations.

A reciprocal relation between potassium and hydrogen is not confined to the renal tubules. There is evidence that when rats and men are depleted of potassium, only two equivalents of sodium enter the muscles for each three equivalents of potassium lost from the cells (Cf. Huth et al., 1959). It has been stated that the balance is made up by one equivalent of hydrogen ions, but it cannot be supposed that the deficit in the intracellular concentration of metallic cations is actually made up by an increase in the concentration of free hydrogen ions. At any possible pH the concentration of free hydrogen ions would be far too small; it could not exceed, say, 0·001 mmol/l at pH 6·0. Hydrogen ions equivalent to one-third of the potassium lost could, however, enter the cells and combine with intracellular buffer anions. The hydrogen ions would not remain free in the intracellular fluid to balance the charges on intracellular anions, but the charges on the buffer anions in the cells would be reduced by the charges on the hydrogen ions which they had taken up.

The decrease in pH of the cell contents which should accompany such a conversion of intracellular buffer bases to their conjugate acids has been demonstrated in the muscles of rats by means of a technique which depends upon the distribution of a non-toxic weak acid between the plasma and the muscles (Saunders et al., 1960; Irvine et al., 1961). If depletion of the body's stores of potassium brought about a similar alteration in the reaction of the renal tubular cells, this should promote the secretion of hydrogen ions and assist in the conservation of potassium. The lowering of intracellular pH could therefore account for the excretion of an acid urine; and the corresponding increase in the rate at which bicarbonate was added to the plasma could account for the alkalosis which is found in potassium-depleted patients who are, paradoxically, excreting an acid urine (Elkinton & Danowski, 1955). The 'alkalosis' is, of course, extracellular; and the acidity of the urine is less paradoxical in relation to the intracellular acidosis which it may be supposed to reflect. An increased output of steroid hormones from the adrenal cortex may be necessary for the (extracellular) renal alkalosis to develop in potassium deficiency (Huth et al., 1959). This would be consistent with the suggestion that adrenal steroids specifically promote the reabsorption of sodium in exchange for other cations (Pitts, 1959). The steroids would stimulate the exchanging process; and the composition of the cell would determine that hydrogen ions would be exchanged rather than potassium.

A lack of reabsorbable anions in a glomerular filtrate from which sodium is avidly reabsorbed is another cause of acidification of the urine (Schwartz et al., 1955) and of addition of bicarbonate to the plasma. In hypochloraemic alkalosis chloride is deficient and its place is taken by bicarbonate which, although it may not be directly reabsorbable, takes up hydrogen ions and is replaced by an equal quantity of bicarbonate in the plasma, thus tending to perpetuate the alkalosis. The alkalosis accompanying depletion of potassium may be only partly due to excessive secretion of hydrogen ions from unduly acid tubular cells. Correction of the alkalosis by giving potassium chloride has been thought to confirm the view that it was due to a cellular deficiency of potassium, but Schwartz et al. (1968) pointed out that it can sometimes be cured with sodium chloride, which cannot correct the deficiency of potassium but does supply chloride to be reabsorbed along with sodium, and may in this way permit the concentration of bicarbonate in the plasma to be reduced.

If a deficiency of potassium in the renal tubular cells can lead to the excretion of an inappropriately acid urine and the addition of too much bicarbonate to the plasma, the administration of potassium salts might be expected to reduce the secretion of hydrogen ions and to promote the excretion of bicarbonate. Potassium salts have in fact long enjoyed a reputation as urinary alkalinizing agents. In an experimental study of this process Black & Mills (1954) administered 50 mmol doses of NaCl, $NaHCO_3$, KCl and $KHCO_3$ respectively to normal human subjects before retiring and collected the urine produced overnight. NaCl had no effect upon the excretion of hydrogen ions; KCl depressed it as much as did $NaHCO_3$, and $KHCO_3$ about twice as much as either of these.

Chapter 26

RENAL ACIDOSIS

A situation somewhat analogous to that described on page 91 in connection with respiratory alkalaemia arises when the secretion of hydrogen ions is depressed, not by a reduction in arterial $P\text{CO}_2$, but by inhibitors of carbonic anhydrase, such as acetazolamide. Bicarbonate is then excreted in the urine (which becomes alkaline) because there is no longer sufficient hydrogen ion secreted to decompose all the filtered bicarbonate. Sodium, with some potassium too, is excreted along with the bicarbonate, so that acetazolamide acts as a diuretic. The loss of bicarbonate leads to a systemic acidosis as the concentration of bicarbonate in the plasma progressively falls. But this process is self-limiting. When the concentration of bicarbonate in the plasma has fallen sufficiently, the reduced rate of secretion of hydrogen ions which can be achieved by the uncatalysed hydration of carbon dioxide again becomes adequate to decompose all the filtered bicarbonate. Then bicarbonate ceases to be excreted, and the diuretic effect fades out.

Inhibitors of carbonic anhydrase produce a systemic acidosis because they interfere with the secretion of hydrogen ions into the tubular urine and the simultaneous addition of corresponding amounts of bicarbonate to the plasma. Pathological processes which interfere with the secretion of hydrogen ions might also be expected to lead to systemic acidosis of renal origin (Elkinton, 1960). Schwartz & Relman (1957) challenged an earlier view that the acidosis which may complicate chronic renal failure was due to the impossibility of eliminating the conjugate bases of acids produced in the body ('acid radicals' as they used to be called) when the rate of glomerular filtration became grossly diminished. They argued that since the excretion of acid depends upon secretion of hydrogen ions, it is essentially a matter for the tubules, and that in fact all renal acidosis must be renal tubular acidosis. This view is in the main correct, but it is a little oversimplified. The performance of the acid-secreting cells depends not only upon the capacity and intensity of their mechanisms for secreting hydrogen ions in exchange for sodium, but also upon the availability of carriers to take hydrogen ions away into the urine. Of the two most important carriers one, ammonia, is produced by the renal tubular cells, and defective formation of ammonia will impair the capacity to secrete hydrogen ions. The other major carrier of hydrogen ions between the secreting cells and the urine is phosphate, and this is delivered to the tubules in the glomerular filtrate, so that its supply depends upon the rate of glomerular filtration. When the rate of filtration is grossly diminished, the rate of excretion of titratable acid may fall below the capacity of the tubular epithelium to secrete hydrogen ions and a state of systemic acidosis may develop, not because, as used to be stated, phosphate is retained in the blood, but because it is not presented to the tubules to be converted into titratable acid.

An elementary consideration of the physiology of the renal excretion of acid suggests three possible causes for systemic acidosis of renal origin:

1. Inadequate metabolic pumping. Deficient secretion of hydrogen ions either in *intensity* (failure to establish a normal difference in pH between urine and blood) or in *quantity*.
2. Insufficiency of buffer in the glomerular filtrate as raw material for making titratable acid.
3. Inadequate formation of ammonia to serve as a carrier of hydrogen ions into the urine as ammonium.

Although the first and the third of these causes should depend mainly upon the number of cells available and upon their metabolic competence, the second should depend upon the rate of glomerular filtration. The formation of ammonia seems to be the most vulnerable function in diseases that destroy nephrons, and the output of ammonium tends to decline more or less in proportion with the reduction in glomerular filtration rate, possibly because glomeruli and tubules are often destroyed in similar numbers. Lack of buffers does not appear to become important until renal failure is far advanced and glomerular filtration severely depressed (Wrong & Davies, 1959).

Chapter 27

SEQUENCE OF EVENTS FOLLOWING PRODUCTION OR ABSORPTION OF ACID

It may be useful to summarize the body's defensive mechanisms by a rough scheme of events following the introduction of strong acid into the blood.

(1) First lines of defence—dilution and buffering.

Acids are diluted rapidly in the extracellular fluid and more slowly in the intracellular fluids of the body. Most of the hydrogen ions disappear by combining with buffer bases, which are converted to their conjugate acids. Some hydrogen ions disappear by exchanging for other cations from bone mineral or from cells. The anions which are the conjugate bases of strong acids whose hydrogen ions have been buffered remain in the place of bicarbonate and other buffer bases which have been used up. The pH of the plasma falls, but falls far less than it would do if dilution did not occur and if the buffers were not present.

(2) Second line of defence—respiratory compensation.

Falling pH promptly stimulates respiration through the peripheral chemoreceptors. When arterial $P\text{CO}_2$ is decreased the pH of the cerebrospinal fluid rises and the intracranial chemoreceptors contribute less to the respiratory drive. Hence when acidosis is of metabolic origin the initial respiratory loss of carbon dioxide causes a paradoxical upward shift in the pH of the cerebrospinal fluid, and the intracranial chemoreceptors check the increased rate of ventilation until the concentration of bicarbonate in the cerebrospinal fluid has been adjusted by mechanisms which remain to be elucidated. Any increase in the rate of pulmonary ventilation removes carbon dioxide faster than before and tends to lower the arterial P_{CO_2} below the normal value of about 40 mmHg. In so far as the concentration of carbonic acid in the plasma is lowered to match the low concentration of bicarbonate, the fall in pH is offset.

(3) Slower restorative measures.

The kidneys generate new hydrogen ions, equivalent to those which have been buffered in the body, and excrete them in the urine (as titratable acids and ammonium) together with the anions which are the conjugate bases of acids to be eliminated. At the same time the kidneys add an equivalent amount of new bicarbonate to the plasma. The addition of this bicarbonate:

(*a*) restores the concentration of bicarbonate in the plasma to a normal value;

(*b*) regenerates other buffers in the body by removing the hydrogen ions which they had taken up (this converts the conjugate acids to buffer bases again);

(*c*) reverses exchanges of ions with cells and with the skeleton as the reaction and the concentration of bicarbonate in the plasma return to normal.

An approximate time scale for such sequential adjustments was indicated by Yoshimura (1965) who administered HCl and $NaHCO_3$ by infusion to dogs. In each case renal excretion was negligible by the time that the infusion was stopped. Most of the acid was buffered in the extracellular compartment (much of it through exchanges with the cells) and most of the $NaHCO_3$ was buffered in the cellular compartment. After 24 hours a quarter of the acid had been excreted, the extracellular fluid was almost normal, and the remaining three-fourths of the acid was being buffered beyond the confines of the inulin space, in bone, muscle, skin and liver in that order. The renal excretion of the remainder of the load of acid occupied about a week.

The dogs given sodium bicarbonate had excreted the excess bicarbonate in 24 hours, but much of the excess sodium was still in the body—most of it outside the extracellular compartment as NaCl. During the next 2 to 7 days this excess NaCl was brought back through the extracellular compartment and excreted.

In summary, the body's most immediate defences are physico-chemical reactions. The first physiological mechanism to operate is the respiratory adjustment of the bicarbonate buffer ratio in two phases, rapid in plasma, slower in the cerebrospinal fluid and probably in many cells. The final regeneration of normal amounts of the components of all the body's buffer systems depends mainly upon the work of the kidneys, and takes much longer than the adjustment of the buffer ratios by the lungs.

REFERENCES

ASMUSSEN E. (1963) The regulation of respiration. In *The Regulation o, Human Respiration*. Ed. D. J. C. Cunningham & B. B. Lloyd, p. 59. Oxford: Blackwell Scientific Publications

ASTRUP P. (1961) A new approach to acid-base metabolism. *Clin. Chem.* **7**, 1

AYLWARD G. H. & FINDLAY T. J. V. (1971) *SI Chemical Data*. Sydney: John Wiley

BARKER E. S. & ELKINTON J. R. (1958) Editorial: Hydrogen ions and buffer bases. *Amer. J. Med.* **25**, 1

BAYLISS L. E. (1959–60) *Principles of General Physiology*. 2 vols. London: Longmans

BELL R. P. (1952) *Acids and Bases. Their Quantitative Behaviour*. London: Methuen

BELL R. P. (1959) *The Proton in Chemistry*. London: Methuen

BERGSTROM W. H. (1956) The skeleton as an electrolyte reservoir. *Metabolism*, **5**, 433

BERLINER R. W. & ORLOFF J. (1956) Carbonic anhydrase inhibitors. *Pharmacol. Rev.* **8**, 137

BLACK D. A. K. & MILLS J. N. (1954) Nocturnal electrolyte excretion after oral administration of sodium and potassium chloride and bicarbonate. *Clin. Sci.*, **13**, 211

BRAZEAU P. & GILMAN A. (1953) Effect of plasma CO_2 tension on renal tubular reabsorption of bicarbonate. *Amer. J. Physiol.* **175**, 33

BROWN E. B. Jr. (1953) Physiological effects of hyperventilation. *Physiol. Rev.*, **33**, 445

BUTLER T. C., WADDELL W. J. & POOLE D. T. (1966) The pH of intracellular water. *Ann. N. Y. Acad. Sci.*, **133**, 73

CAMPBELL E. J. M. (1962) RI pH. *Lancet*, **i**, 681; **ii**, 154

CAMPBELL E. J. M. (1968) Hydrogen ion (acid-base) regulation. In *Clinical Physiology*. 3rd Edn. Eds. E. J. M. Campbell, C. J. Dickinson & J. D. H. Slater, Chap. 5, p. 202. Oxford: Blackwell Scientific Publications

CARLSON L. D. (1960) Gas exchanges and transportation. In *Medical Physiology and Biophysics*. Eds. T. C. Ruch & J. F. Fulton. 18th Edn. Chap. 36, p. 789. Philadelphia & London: W. B. Saunders Co.

CARTER N. W., RECTOR F. C. & SELDIN D. W. (1966) Relation between intracellular pH and membrane potential in rat skeletal muscle. *III Int. Congr. Nephrology Abstracts*, 170

CHINARD F. P. (1966) An elementary approach to the carbon dioxide system. *Ann. N. Y. Acad. Sci.*, **133**, 87

CLAPP J. R., RECTOR F. C., JR. & SELDIN D. W. (1962) Effect of unabsorbed anions on proximal and distal tubular potentials in rats. *Amer. J. Physiol.*, **202**, 781

CONWAY E. J. (1957) Nature and significance of concentration relations of potassium and sodium ions in skeletal muscle. *Physiol. Rev.*, **37**, 84

COTTON F. A. & WILKINSON G. (1962) *Advanced Inorganic Chemistry. A comprehensive text*. Interscience

CREESE R., NEIL M. W., LEDINGHAM J. M. & VERE D. W. (1962) The terminology of acid-base regulation. *Lancet*, **i**, 419

DAVENPORT H. W. (1958) *The ABC of Acid-Base Chemistry*. 4th Edn. University of Chicago Press

DAVIES H. W., HALDANE J. B. S. & KENNAWAY E. L. (1920) Experiments on the regulation of the blood's alkalinity. *J. Physiol.*, **54**, 32

DAVIES R. E. (1951) The mechanism of hydrochloric acid production by the stomach. *Biol. Rev.*, **26**, 87

DAVIS R. P. (1967) Logland: A Gibbsian view of acid-base balance. *Amer. J. Med.*, **42**, 159

DORMAN P. J., SULLVIAN W. J. & PITTS R. F. (1954) The renal response to acute respiratory acidosis. *J. clin. Invest.*, **33**, 82

EDSALL J. T. & WYMAN J. (1958) *Biophysical Chemistry*, Vol. I. London & New York: Academic Press

ELKINTON J. R. (1956) Whole body buffers in the regulation of acid-base equilibrium. *Yale J. Biol. & Med.*, **29**, 191

ELKINTON J. R. (1960) Editorial: Renal acidosis. *Amer. J. Med.*, **28**, 165

ELKINTON J. R. & DANOWSKI T. S. (1955) *The Body Fluids, Basic Physiology and Practical Therapeutics*. Baltimore: Williams & Wilkins

FILLEY G. F. (1971) *Acid-base and Blood Gas Regulation*. Philadelphia: Lea & Febiger

FOURMAN P. (1960) *Calcium Metabolism and the Bone*. Oxford: Blackwell Scientific Publications

FRAZER S. C. & STEWART C. P. (1959) Acidosis and alkalosis, a modern view. *J. clin. Path.*, **12**, 195

GAMBINO S. R. (1966) General discussion. *Ann. N. Y. Acad. Sci.*, **133**, 244

GIEBISCH G., BOULPAEP E. L. & WHITTEMBURY G. (1971) Electrolyte transport in kidney tubule cells. *Philos. Trans. Roy. Soc. Lond.*, **B, 262,** 175

GIEBISCH G. & MALNIC G. (1970) Some aspects of renal tubular hydrogen ion transport. *Proc. int. Congr. Nephrol.* Washington, 1969. Vol. I, p. 181

GIEBISCH G. & WINDHAGER E. E. (1964) Renal tubular transfer of sodium chloride and potassium. *Amer. J. Med.*, **36**, 643

GILMAN A. (1958) The mechanism of diuretic action of the carbonic anhydrase inhibitors. *Ann. N. Y. Acad. Sci.*, **71**, 355

GOODMAN A. D. & FUISZ R. E. (1964) Mechanism of regulation of renal bicarbonate reabsorption by plasma CO_2 tension. *Amer. J. Physiol.*, **206**, 719

GOTTSCHALK C. W., LASSITER W. E. & MYLLE M. (1960) Localization of urine acidification in the mammalian kidney. *Amer. J. Physiol.*, **198**, 581

GRAY J. S. (1949) *Pulmonary Ventilation and its Physiological Regulation.* Springfield, Ill.: Charles C Thomas

GUGGENHEIM E. A. (1957) *Thermodynamics, an Advanced Treatment for Chemists and Physicists.* Amsterdam: North-Holland Publishing Co.

HALDANE J. B. S. (1921) Experiments on the regulation of the blood's alkalinity. *J. Physiol.*, **55**, 265

HALDANE J. S. & PRIESTLEY J. G. (1935) *Respiration.* New Edition. Oxford: University Press

HASTINGS A. B. (1966) Introductory remarks. *Ann. N. Y. Acad. Sci.*, **133**, 15

HENDERSON L. J. (1913) *The Fitness of the Environment.* New York, Macmillan

HUNT J. N. (1956) The influence of dietary sulphur on the urinary output of acid in man. *Clin. Sci.*, **15**, 119

HUTH E. J., SQUIRES R. D. & ELKINTON J. R. (1959) Experimental potassium depletion in normal human subjects. II. Renal and hormonal factors in the development of extracellular alkalosis during depletion. *J. clin. Invest.*, **38**, 1149

IRVINE R. O. H., SAUNDERS S. J., MILNE M. D. & CRAWFORD M. A. (1961) Gradients of potassium and hydrogen ion in potassium deficient voluntary muscle. *Clin. Sci.*, **20**, 1

ITO M., KOSTYUK P. G. & OSHIMA T. (1962) Further study on anion permeability of inhibitory post-synaptic membrane of cat motoneurones. *J. Physiol.* **164**, 150

LEAF A. (1960) Kidney, water and electrolytes. *Ann. Rev. Physiol.*, **22**, 111

LEMANN J. Jr. & RELMAN A. S. (1959) The relation of sulfur metabolism to acid-base balance and electrolyte excretion. *J. clin. Invest.*, **38**, 2215

LEUSEN (1972) Regulation of cerebrospinal fluid composition with reference to breathing. *Physiol. Rev.*, **52**, 1

LOTSPEICH W. D. (1959) *Metabolic Aspects of Renal Function.* Springfield, Ill.: Charles C Thomas

MCGLASHAN M. L. (1968) *Physicochemical Quantities and Units.* London: Royal Institute of Chemistry.

MacKnight A. D. C., MacKnight J. M. & Robinson J. R. (1962) The effect of urinary output upon the excretion of ammonia in man. *J. Physiol.*, **163**, 314

Malnic G. & Aires M. (1970) Micropuncture study of anion transfer in proximal tubules of rat kidney. *Amer. J. Physiol.*, **218**, 27

Malvin R. L., Wilde W. S. & Sullivan L. P. (1958) Bicarbonate reabsorption along renal tubules. *Proc. Soc. exp. Biol. Med.*, **98**, 448

Maren T. H. (1969) Renal carbonic anhydrase and the pharmacology of sulfonamide inhibitors. *Handb. expt. Pharmakol.*, **24**, 195

Milne M. D., Scribner B. H. & Crawford M. A. (1958) Non-ionic diffusion and the excretion of weak acids and bases. *Amer. J. Med.*, **24**, 709

Mitchell R. A. (1966) Cerebrospinal fluid and the regulation of respiration. In *Advances in Respiratory Physiology*. Ed. C. G. Caro, p. 1. London: Edward Arnold

Nahas G. G. (Ed.) (1966) Current concepts of acid-base measurement. *Ann. N. Y. Acad. Sci.*, **133**, Art. 1

Nichols G. Jr. & Nichols N. (1956) The role of bone in sodium metabolism. *Metabolism*, **5**, 438

Ochwadt B. K. & Pitts R. F. (1956) Effects of intravenous infusion of carbonic anhydrase on carbon dioxide tension of alkaline urine. *Amer. J. Physiol.*, **185**, 426

Pappenheimer J. R. (1967) The ionic composition of cerebral extracellular fluid and its relation to control of breathing. *Harvey Lect*, **61**, 71

Parsons T. R. (1920) The reaction and carbon dioxide carrying power of the blood. A mathematical treatment. Part II, *J. Physiol.*, **53**, 340

Peters J. P. & van Slyke D. D. (1931) Quantitative clinical chemistry. Vol. I. *Interpretations*. London: Ballière, Tindall & Cox

Pitts R. F. (1950) Acid-base regulation by the kidneys. *Amer. J. Med.*, **9**, 356

Pitts R. F. (1953) Mechanisms for stabilizing the alkaline reserves of the body. *Harvey Lect.*, **48**, 172

Pitts R. F. (1959) *The Physiological Basis of Diuretic Therapy*. Springfield, Ill.: Charles C Thomas

Pitts R. F. (1968) *Physiology of the Kidney and Body Fluids*. 2nd Edn. Chicago: Year Book Medical Publishers

Pitts R. F. (1964) Renal production and excretion of ammonia. *Amer. J. Med.*, **36**, 720

Pitts R. F. (1971) The role of ammonia production and excretion in regulation of acid base balance. *New Eng. J. Med.*, **284**, 32

PITTS R. F. & ALEXANDER R. S. (1945) The nature of the renal tubular mechanism for acidifying the urine. *Amer. J. Physiol.*, **144**, 239

PITTS R. F., GURD R. S., KESSLER R. H. & HIERHOLZER K. (1958) Localization of acidification of urine, potassium and ammonia secretion and phosphate reabsorption in the nephron of the dog. *Amer. J. Physiol.*, **194**, 125

PITTS R. F. & LOTSPEICH W. D. (1946) Bicarbonate and the renal regulation of acid-base balance. *Amer. J. Physiol.*, **147**, 138

PITTS R. F., SULLIVAN W. J. & DORMAN P. J. (1954) Regulation of the content of bicarbonate bound base in body fluids. In CIBA Foundation Symposium on *The Kidney*. Ed. A. A. G. Lewis & G. E. W. Wolstenholme, p. 125. London: J. & A. Churchill, Ltd.

RECTOR F. C. Jr., CARTER N. W. & SELDIN D. W. (1965) The mechanism of bicarbonate reabsorption in the proximal and distal tubule of the kidney. *J. clin. Invest.*, **44**, 278

RICHET G., ARDILLOU R. & AMIEL C. (1967) Les phospholipides, source alimentaire d'ions H^+. *Proc. 3rd int. Congr. Nephrol.* Washington, 1966, Vol. I, p. 136

RICHTERICH R. (1962) Physicochemical factors determining ammonia excretion. *Helv. Physiol. Acta* **20**, 236

ROBIN E. O. & BROMBERG P. A. (1959) Editorial: Claude Bernard's milieu intérieur extended: intracellular acid-base relationships. *Amer. J. Med.*, **27**, 689

ROBSON J. S., BONE J. M. & LAMBIE A. T. (1968) Intracellular pH. *Clin. Chem.*, **11**, 213

SAUNDERS S. J., IRVINE R. O. H., CRAWFORD M. A. & MILNE M. D. (1960) Intracellular pH of potassium-deficient voluntary muscle. *Lancet*, **i**, 468

SCHWARTZ W. B., JENSON R. L. & RELMAN A. S. (1955) Acidification of the urine and increased ammonium excretion without change in acid-base equilibrium: sodium reabsorption as a stimulus to the acidifying process. *J. clin. Invest.*, **34**, 673

SCHWARTZ W. B. & RELMAN A. S. (1957) Acidosis in renal disease. *New Engl. J. Med.*, **256**, 1184

SCHWARTZ W. B., VAN YPERSELE DE STRIHOU C. & KASSIRER J. P. (1968) The role of anions in metabolic alkalosis and potassium deficiency. *New Engl. J. Med.*, **279**, 630

SELDIN D. W., PORTWOOD R. M., RECTOR F. C. Jr. & CADE R. (1959) Characteristics of renal bicarbonate reabsorption in man. *J. clin. Invest.*, **38**, 1663

Severinghaus J. W., Mitchell R. A., Richardson B. W. & Singer M. M. (1963) Respiratory control at high altitude suggesting active transport regulation of CSF pH. *J. Appl. Physiol.*, **18**, 1155

Siggaard Andersen O., Engel K., Jørgensen K. & Astrup P. (1960) A micromethod for determination of pH, carbon dioxide tension, base excess and standard bicarbonate in capillary blood. *Scandinav. J. clin. lab. Invest.*, **12**, 172

Singer R. B., Clark J. K., Barker E. S., Crosley A. P. Jr. & Elkinton J. R. (1955) The acute effects in man of rapid intravenous infusion of hypertonic sodium bicarbonate solution. I. Changes in acid-base balance and distribution of the excess buffer. *Medicine*, **24**, 51

Singer R. B. & Hastings A. B. (1948) An improved clinical method for the estimation of disturbances of the acid-base balance of human blood. *Medicine*, **27**, 223

Smith H. W. (1951) *The Kidney. Structure and Function in Health and Disease*. New York: Oxford Medical Publications

Smith H. W. (1956) *Principles of Renal Physiology*. New York: Oxford University Press

Swan R. C., Pitts R. F. & Madisso H. (1955) Neutralization of infused acid by nephrectomized dogs. *J. clin. Invest.*, **34**, 205

Ullrich K. J., Eigler F. W. & Pehling G. (1958, a) Sekretion von Wasserstoffionen in den Sammelrohren der Säugetierniere. *Pflugers Archiv.*, **267**, 491

Ullrich K. J., Hilger H. H. & Klumper J. D. (1958, b) Sekretion von Ammoniumionen in den Sammelrohen der Säugetierniere. *Pflugers Archiv.*, **267**, 244

Van Slyke D. D. (1921) Studies of acidosis. XVII. The normal and abnormal variations in the acid-base balance of the blood. *J. biol. Chem.*, **48**, 153

Waddell W. J. & Bates R. G. (1969) Intracellular pH. *Physiol. Rev.* **49**, 285

Waddell W. J. & Butler T. C. (1959) Calculation of intracellular pH from the distribution of 5,5-dimethyl-2,4-oxazolidinedione (DMO). Application to skeletal muscle of the dog. *J. clin. Invest.*, **38**, 720

Walker A. M., Bott P. H., Oliver J. & MacDowell M. C. (1941) The collection and analysis of fluid from single nephrons of the mammalian kidney. *Amer. J. Physiol.*, **134**, 580

Walker W. D., Goodwin F. J. & Cohen R. D. (1969) Mean intracellular hydrogen ion activity in the whole body, liver, heart and skeletal muscle of the rat. *Clin. Sci.*, **36**, 409

WEINER I. M. & MUDGE G. H. (1964) Renal tubular mechanisms for excretion of organic acids and bases. *Amer. J. Med.*, **36**, 743

WHITE A., HANDLER P., SMITH E. L. & STETTEN DEWITT Jr. (1959) *Principles of Biochemistry*. 2nd Edn. New York & London: McGraw-Hill Book Co. Inc.

WICK T. & FROMTER E. (1967) Das Zellpotential des proximalen Konvoluts der Rattenniere in Abhängigkeit von der peritubulären lonenkonzentration. *Pflugers Archiv.*, **294**, R17

WINDHAGER E. E. (1968) *Micropuncture Techniques and Nephron Function*. London: Butterworths

WINTERS R. W. & DELL R. B. (1965) Regulation of acid-base equilibrium. Chap. 10, p. 181. In *Physiological Controls and Regulations*. Eds. W. S. Yamamoto & J. R. Brobeck. Philadelphia & London: W. B. Saunders Co.

WOODBURY J. W. (1965) Regulation of pH. Chap. 46, p. 899. In *Physiology and Biophysics*. 19th Edn. Eds. T. C. Ruch & H. D. Patton. Philadelphia & London: W. B. Saunders Co

WRONG O. & DAVIES H. E. F. (1959) The excretion of acid in renal disease, *Quart. J. Med.*, **28**, 259

WYNN V. (1959) The clinical significance of blood pH and blood gas measurement. In *A Symposium on pH and Blood Gas Measurement*. Ed. R. F. Woolmer, p. 166. London: J. & A. Churchill, Ltd.

YOSHIMURA H. (1966) Tissue buffering and control of acid-base balance in body fluids. *Proc. I.U.P.S. IV Int. Congr., Tokyo*, 189

INDEX